THE
MARINE ENVIRONMENT
AND
STRUCTURAL DESIGN

THE
MARINE ENVIRONMENT
AND
STRUCTURAL DESIGN

John Gaythwaite, P.E.
Consulting Engineer

VNR **VAN NOSTRAND REINHOLD COMPANY**
NEW YORK   CINCINNATI   ATLANTA   DALLAS   SAN FRANCISCO
LONDON   TORONTO   MELBOURNE

Van Nostrand Reinhold Company Regional Offices:
New York   Cincinnati   Atlanta   Dallas   San Francisco

Van Nostrand Reinhold Company International Offices:
London   Toronto   Melbourne

Library of Congress Catalog Card Number: 80-29583
ISBN: 0-442-24834-2

Manufactured in the United States of America

Published by Van Nostrand Reinhold Company
135 West 50th Street, New York, N.Y. 10020

Published simultaneously in Canada by Van Nostrand Reinhold Ltd.

15 14 13 12 11 10 9 8 7 6 5 4 3 2 1

Library of Congress Cataloging in Publication Data

Gaythwaite, John.
   The marine environment and structural design.

   Includes bibliographical references and index.
   1. Ocean engineering. 2. Structural design.
I. Title.
TC1645.G39   627   80-29583
ISBN 0-442-24834-2

To My Parents
John I. and Doris Gaythwaite

Preface

In my formative years as a waterfront engineer, I spent a good deal of time seeking information on the ocean and its physical behavior. There were textbooks on oceanography, reference works on harbor construction, and here and there technical papers from various journals that attempted to relate some aspects of ocean science to structural design. However, I was seeking a single reference that tied it all together and presented an overview of the marine environment as it affects structures in the sea. Since I never could find such a reference, I decided to write one myself—hence this compilation of information on the sea as viewed by a structural engineer. My primary objective is to give those trained in the traditional disciplines of civil and structural engineering and perhaps in naval architecture a background in the nuances of the marine environment that may aid them in becoming specialists in waterfront, coastal, or the broader realm of ocean engineering. I hope it also will be useful as a supplementary text for introductory courses in coastal, harbor, and ocean structures engineering. The introductory chapter elaborates further on the scope and purpose of this book.

I believe the book is unique in attempting to cover all aspects of the marine environment that particularly relate to the design of structures constructed in and around the sea. Further, in the past decade much has been written on many aspects of the development of the oceans, in subject areas where information was hitherto quite sparse. I trust that this book will find its place among the growing body of literature in this important and exciting area of human knowledge.

I am greatly indebted to those who over many years have generated the body of knowledge that I have attempted to bring together in this book. I have tried to cite as many of the more outstanding and useful works as possible, realizing that even the most thorough and meticulous search of the literature is bound to miss some important contributions, as well as the original contributors of some bits of information. I am additionally indebted to those who have encouraged and influenced the development of my career as a professional engineer, in particular Mr. Paul S. Crandall, whose devotion and enthusiasm are inspirational and in whose employ (Crandall Dry Dock Engineers, Inc.) I gained important experience and knowledge of the waterfront. As regards this book, I especially wish to express my gratitude to Dr. Gerard Mangone, director of the Center for the Study of Marine Policy, University of Delaware, for his encouragement and helpful comments on the manuscript, to Mrs. Beatrice Richards of the W. F. Clapp Laboratories for reviewing the material on marine borers and fouling, to Mrs. Margaret Ruth Rich for typing all the final manuscript, and finally to my wife, Michele Rabot Gaythwaite, for her assistance in preparation of the manuscript and her patience throughout the book's development.

John W. Gaythwaite

Manchester, Massachusetts

Notations

A = area, cross-sectional area
A_0 = area of virtual mass
A_p = projected area
A_s = astronomical tide height at time of storm surge
a = wave amplitude = half wave height
B = vessel beam
b = orthogonal spacing
 = width
C = wave celerity
C_b = breaking wave force coefficient
C_D = drag force coefficient
C_F = friction drag coefficient
C_G = wave group celerity
C_i = ice force coefficient
C_L = lift force coefficient
C_M = inertial force coefficient
C_0 = initial or deep water wave celerity
C_R = wave reflection coefficient
C_T = wave transmission coefficient
c = coefficient for ice strength equation
D = pile diameter
 = vessel draft
D.W.L. = design water level
d = water depth
E = energy
E_h = oxidation-reduction potential
E_k = kinetic energy
E_p = potential energy
e = base of Naperian logarithms
F = fetch length
F_D = drag force
F_c = current force

F_{dm} = maximum wave drag force
F_i = inertial force
F_{im} = maximum wave inertial force
F_L = lift force
F_{Tm} = maximum total force
f = frequency = $1/T$
 = friction factor
f_{ci} = ice compressive strength
G_z = wind gust velocity at elevation Z
G_{30} = wind gust velocity at 30 feet above surface
g = acceleration of gravity at earth's surface
H = wave height
H_b = height of breaking wave
H_0 = initial or deep water wave height
H_s = significant wave height
$\bar{H}, H_{RMS}$ = root mean square wave height
$H_{1/10}$ = average of highest 10% of waves in record
K_D = wave diffraction coefficient
K_{dm} = wave drag force factor
K_{FP} = friction-percolation coefficient
K_{im} = wave inertial force factor
K_R = wave refraction coefficient
K_S = shoaling coefficient
K_1 = soli-lunar tide-producing component
k = wave number = $2\pi/L$
k_c = Kuelegan-Carpenter number
L = wavelength
L_A = Airy wavelength
L_c = wavelength on current
L_0 = initial or deep water wave length

L.W.L. = length along vessel's waterline

M = mass
= parameter in solitary wave
equation

M' = virtual mass

M_{dm} = moment due to maximum
wave drag force

M_{im} = moment due to maximum
wave inertial force

M.H.W. = mean high water

M.L.W. = mean low water

M.S.L. = mean sea level

M.T.L. = mean tide level

M_2 = principal lunar tide-producing
component

N = parameter in solitary wave
equation

N_F = Froude number

N_R = Reynolds number

N_S = Strouhal number

n = wave transmission coefficient
= an integer number

0_1 = principal lunar diurnal tide-
producing component

P = force per unit area
= power

$P_{(H)}$ = probability of occurrence
of H

P_{max} = maximum wave pressure

PMH = probable maximum hurricane

P_S = pressure setup

pcf = pounds per cubic foot

pH = negative log of hydrogen ion
concentration (measure of
acidity)

psf = pounds per square foot

Q = volume of wave overtopping

R = wave run-up height
= radius to extreme winds
= radius of trochoidal wave generating circle

R_T = return period
= tide range

r = radius vector
= radius of wave particle orbit

r_0 = radius of wave particle orbit at
surface

S = surface area or wetted surface area

SCE = saturated calomel electrode

S_d = level arm for wave drag force

S_i = lever arm for wave inertial force

SPH = standard project hurricane

STP = standard temperature and
pressure

S.W.L. = still water level

$S(\omega)$ = wave spectral density

S_1 = principal solar tide producing
component

T = wave period

T_n = natural period of basin or
structure

T_0 = initial or deep water wave
period

T_S = significant wave period

t = duration
= time

U = current velocity

U_s = surface current velocity

U_z = velocity of current at elevation Z

u = velocity component

u_{max} = maximum wave particle
velocity at crest

V = wind velocity

V_{cr} = critical velocity

V_F = storm forward speed

V_R = velocity at distance R from
storm center

V_z = wind velocity at elevation Z

V_1 = one-minute sustained wind
speed

V_{30} = wind velocity at 30 feet above
surface

v = velocity component

W_s = wind set-up

W_w = wave set-up

w = velocity component

x, y, z = coordinate directions

α = storm speed correction factor
  = angle between wave crest and bottom contour
  = wave particle acceleration
β = angle (as defined in given equations)
γ = unit weight or specific weight
ΔP = drop in atmospheric pressure at storm center
ΔS = incremental distance over which depth remains constant
ΔX = incremental distance
ζ = vertical wave particle displacement
η = wave surface profile
η_c = wave crest elevation above S.W.L.
θ = phase angle = $(kx - \omega t)$
  = angle as defined in given equation

λ = aspect ratio
μ_e = eddy viscosity
ν = kinematic viscosity
  = brine volume
ξ = horizontal wave particle displacement
ρ = mass density = γ/g
σ = ice strength
σ_t = seawater density function
T_B = bottom shear stress
T_s = surface shear stress
T_ν = wave decay period
ϕ = solidity ratio
  = velocity potential
  = latitude
Ω = angular velocity of earth
ω = angular frequency = $2\pi/T$

Abbreviations for Professional Societies and Organizations

ANSI	American National Standards Institute
API	American Petroleum Institute
ASCE	American Society of Civil Engineers
ASME	American Society of Mechanical Engineers
BEB	Beach Erosion Board (U. S. Army)
CRREL	Cold Regions Research and Engineering Laboratory (U. S. Army)
MTS	Marine Technology Society
NACE	National Association of Corrosion Engineers
NAVFAC	Naval Facilities Engineering Command (U. S. Navy)
NOAA	National Oceanic and Atmospheric Administration (U. S. Department of Commerce)
NOS	National Ocean Survey (division of NOAA)
OCIMF	Oil Companies International Marine Forum
OTC	Offshore Technology Conference
PIANC	Permanent International Association of Navigational Congresses
SNAME	Society of Naval Architects and Marine Engineers
USACE	U. S. Army Corps of Engineers
USCGS	U. S. Coast and Geodetic Survey (U. S. Department of Commerce)
USNCEL	U. S. Naval Civil Engineering Laboratory
USNOO	U. S. Naval Oceanographic Office
WMO	World Meteorological Organization

Contents

THE
MARINE ENVIRONMENT
AND
STRUCTURAL DESIGN

1
Introduction

Artists and poets, sailors and explorers, merchants and treasure hunters, since the dawn of civilization all have come down to the water's edge, to gaze in wonder at the majesty of the sea, in all its vicissitudes, its relentless and unpredictable moods, and the myriad life forms above and below its restless surface. Scientists and engineers have come also, to study it, measure it, understand it, and build upon it, in order to coexist with that which was their beginning and which continues to maintain the balance of life on planet Earth. Today we must recognize the crucial nature of that delicate balance as we exploit the sea as a source of food, energy, and raw materials, and as a medium of transportation and recreation. The design of civil engineering structures is an important part of coexistence with the sea, serving as a foothold for further exploitation of the ocean's vast wealth. Those who plan, design, and construct such structures—fixed or floating, at the water's edge or far from sight of land—must understand something of the ocean's moods; they must try to predict its motions, in order that they can design their structures not only to withstand the ravages of nature but to live in harmony with it, in harmony with the many interacting physical, chemical, and biological processes going on about them.

1.1 SCOPE AND PURPOSE

The primary purpose of this book is to give the reader an overview of the marine environment as it affects the design of structures in the sea. The book is oriented toward contemporary civil engineering structures, but as most ocean engineering projects are interdisciplinary in nature, it is believed that much of the information should be of general interest. In particular, the information presented in this book has been drawn from the literature in the fields of civil engineering, naval architecture and applied (physical) oceanography. The title ocean engineer came into vogue in the 1970s. An ocean engineer may be an engineer trained in any of the basic disciplines, but one who also has some knowledge of the ocean environment as it relates to his or her field of endeavor. Although this book is oriented towards ocean/structural engineers, it presents basic information on the marine environment that should concern engineers of any discipline. In the regime of offshore structures and various special-purpose floating structures, there has necessarily been some blending of the traditional disciplines of naval architecture and civil engineering.

This text is intended to cover all aspects of the marine environment as they pertain to structural design, particularly as they affect design loadings, and the selection of materials. Wind, waves, currents, and ice constitute the major sources of environmental loads, while tides and water level variations affect function and operation, and corrosion, fouling, and attack by marine organisms affect selection of materials and the structures' durability and longevity. Figure 1.1 from reference 1 is a pictorial summary of the environmental considerations elaborated

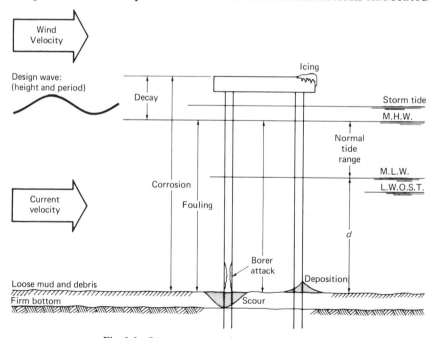

Fig. 1.1. Ocean structure–design considerations.

on in this text. Submarine soil mechanics and siesmic loadings are not covered because of the scope and depth of these subject areas and the fact that these subjects have been well treated elsewhere.

The marine environment presents a wide range of problems. However, information relevant to a particular problem may be difficult to find, it may be hidden in some obscure technical journal, or it may not exist at all. This book necessarily draws heavily upon the work of others, and I have tried to include references to all of the more important works that have useful application to structural design problems. Theoretical derivations and mathematical manipulations have been avoided so that the subject matter should flow smoothly. However, thorough understanding and application of the concepts presented involve a firm grasp of engineering mathematics and statistics on the intermediate to advanced levels, as well as some background in hydrodynamics and fluid mechanics. The book is not intended to be a design guide, but rather to serve as review of environmental factors that affect design criteria.

Two important aspects of designing marine structures should become apparent: one is the importance of the judicious selection of design criteria through an understanding of the nature of environmental loadings and a consideration of the probability of occurence of extreme events versus economics and safety; and the other is the importance of the proper selection and utilization of materials with regard to the structure's function and longevity. Chapters 2 through 6 deal with environmental sources of loadings, whereas Chapter 7 discusses some environmental factors affecting durability of materials.

The structural designer obviously cannot be an expert in all the subject areas discussed, but he/she should be sufficiently informed in these areas to interact with specialists from other fields in order to determine loading and functional design criteria. The designer must always be something of an investigator who can delve into the literature and determine how to go about obtaining sufficient knowledge for the particular problem at hand. He/she must be eclectic in applying the latest information available from the various fields of study. This book is intended to fill the gap between the structural engineer and the oceanographer and to provide sufficient references and reference material that answers can be found to a wide variety of marine design problems. The importance of understanding basics of other, related fields of study cannot be overemphasized. Ultimately there is no substitute for sound engineering judgment based upon the scientific method of data collection and analysis and the engineering approach to problem solving.

A brief note about mathematical units is warranted. No attempt has been made herein to make use of entirely consistent units, such as the SI system; instead I present information in the units in which it is usually encountered in practice in the United States, and in the case of data reproduced from other sources the units of the original presentation have been retained. Moreover, in addition to metric and English units, marine work utilizes some units of its own (e.g., nautical miles, knots, fathoms, etc.). The ocean engineer should become accustomed to thinking in different units because he/she will have to deal with data and dimensions from many sources. Unfortunately a unified system does not yet exist in this country, and probably will not for some time to come. Some important conversions are included in the appendixes at the end of this book.

1.2 HISTORICAL PERSPECTIVE

The development of ports and harbors dates back to antiquity. The earliest documented port and harbor works are attributed to the early Egyptians and Phoenicians, although there is reason to believe that artificial harbors in some embryonic form existed even before that time, during the earliest Eastern civilizations, for which there are no written records.[2] It is well known that both Egypt and Phoenicia had commercial navies and maintained a relatively elaborate system of trading operations. The port of A-UR existed on a branch of the Nile River about 3000 B.C. Artificial harbor works existed also at the ancient cities of Tyre and Sidon, which included quays and warehouse structures. The Port of Pharos at Alexandria in 2000 B.C. was perhaps one of the most striking of early marine construction projects. The harbor works included an 8500-foot-long rubble mound breakwater and later the famous Pharos lighthouse, built around 270 B.C. The Greeks constructed notable harbor works at Rhodes, Salamis, Corinth, Syracuse, and Piraeus.

The Romans possessed a certain genius and capacity for harbor construction. The remains of some of their works still exist today, such as the port works at Ostia, now 20 miles inland from the mouth of the Tiber River because of an advancing shoreline. The Romans developed techniques for pile driving, constructed cofferdams, and developed a hydraulic cement called pozzuolana with which they built concrete seawalls by about 300 B.C. The early Roman port of Civita Vecchia still possesses a serviceable harbor capable of receiving vessels of up to 20-foot draft.[3]

Medieval times and then the discovery of the New World saw a vast increase in the number of vessels and the variety of cargoes; but the design and construction of port and harbor works remained much more of an art based on past experience than a science with any degree of understanding of the natural forces involved. The Industrial Revolution in England brought about some early wave theories and empirical formulas for design. In 1809 Gerstner[4] presented his trochoidal wave theory. The small amplitude linear wave theory was presented by Airy[5] in 1845, and shortly thereafter Stokes[6] presented his work on oscillatory waves. These early theories served as a foundation for contemporary wave theories and analysis techniques. However, it was not until World War II and its crucial amphibious operations that scientists and engineers truly turned their attention to applied ocean science.

In 1938 the first offshore oil rig was constructed by Humble Oil Co. in 60 feet of water, approximately one mile offshore in the Gulf of Mexico. It was founded on timber piles and connected to shore by a timber trestle. Earlier platforms had been constructed in Lake Maracaibo in the 1920s, and the world's first overwater oil well was drilled in Coddo Lake near Shreveport, Louisiana in 1911. It consisted of a wooden platform founded on cypress piles in only 10 feet of water.[7]

The forerunner of today's monumental offshore platforms was constructed in 1947 for the Kerr-McGee Co. in 20 feet of water, 12 miles offshore in the Gulf of Mexico (see Fig. 1.2). It was a jacket-template type platform first conceived by M. B. Willey of J. Ray McDermott & Co., and consisted of tubular steel members connected by bracing. The platform structure was prefabricated and towed to the site. After positioning, steel pipe piles were driven through the hollow columns (jackets). This method of construction is still widely used today. In January of 1961 the tragic failure of Texas Tower No.

Fig. 1.2. Kerr-Mc Gee Offshore Oil Platform, Ship Shoal, Block 32, as it appears today (1977). The platform, built in 1947, was the site of the world's first commercial production of oil from an offshore well out of sight of land. The platform is located about 45 miles south-southwest of Morgan City, La. in the Gulf of Mexico. The first well is still producing oil. (*Photo Courtesy of Kerr-McGee Corp.*)

4, a U. S. Air Force radar platform, during a winter storm off the New Jersey coast, pointed out the need for better understanding of structure and wave interaction and precipitated studies of tower dynamics in random waves. Today, fixed platform structures stand in up to 1000-foot water depths (see Fig. 1.3), located at the very edge of the continental shelves. They are constructed in harsh environments such as the Gulf of Alaska and the North Sea. There are numerous structural configurations, both of steel and of concrete.

However, it has not been just in the offshore oil industry that marine structures have made vast strides forward within the last two decades. The development of offshore and onshore ports and terminals, spurred by advances in materials handling techniques, has also shown the application of innovative and sophisticated applied science. Deeper-draft and larger vessels have generated the

Fig. 1.3. Worlds tallest fixed offshore platform, completed 1979, stands in 1025 feet of water twelve miles south of the mouth of the Mississippi River. The platform weighs 59,000 tons and has 62 wellslots. (*Photo Courtesy of Shell Oil Co.*)

development of new concepts in vessel mooring and cargo handling, such as the unloading of tankers at an offshore single point mooring buoy, which also conveys the liquid cargo from ship to shore via flexible hoses and underwater pipelines. The development of new vessel types such as container ships, Lash, and Roll-on-roll off (Ro-Ro) ships has been accompanied by new developments in wharf design. (Other contemporary marine structures are discussed further in Section 1.4 of this chapter.)

It is apparent from the foregoing discussion that although the design of marine structures has its roots in ancient history, it has only very recently come into its own as an applied science.

(Inman[8] has reviewed the history of harbor engineering, and Bruun[9] the history of coastal defense works, and the interested reader is referred to their work for more information of a historical nature.)

1.3 THE MARINE ENVIRONMENT

The marine environment presents a most harsh and challenging setting for the structural engineer. Witness the Gulf of Mexico, where structures must be designed to withstand hurricane winds of perhaps 150 knots or more with 50- to 70-foot-high waves, while the typically soft bottom sediments present difficult foundation problems, and the persistently hot, humid climate makes for high rates of corrosion of metal structures. Consider the problems of designing an offshore structure in Cook Inlet, Alaska, where tidal currents may move masses of ice at speeds up to 9 knots, where the twice-daily tide range is 30 feet, and high winds, below-freezing temperatures, and large ice cover are the norm; or perhaps the North Sea, where 30- to 40-foot waves are an annual occurence, and 65-foot-high waves occur on an average of every 50 years. Sea structures are exposed to the full fury of natural events and meteorological extremes.

The seas are in constant motion, driven primarily by the natural forces of atmospheric and geological disturbances, density differences of water masses, and astronomical influences such as the sun and moon, all of which are modified by the rotation of the earth and the boundary conditions imposed by the land masses, sea bed, and atmosphere. We can group these motions under the general headings of waves, tides, and currents.

Waves are periodic undulations of the sea surface, although internal waves may also exist deep in the ocean at the interface of water masses of different density. Most waves (which are discussed in some detail in Chapter 3) are generated by the wind. Waves can also be impulsively generated by such phenomena as earthquakes, landslides, and explosions. Figure 1.4, from reference 10, is a composite period spectrum representing the typical ranges in periods (time between the passage of two crests) of periodic changes in elevation of the sea surface. Wind waves, which respond to the restoring forces of gravity and viscosity, typically range in period from 1 to 20 seconds, while most impulsively generated waves such as earthquake waves, called tsunamis (see Chapter 5), range in period from minutes to hours and thus are considered long-period waves. Capillary waves, which are governed by surface tension, are of extremely short periods and amplitude. Capillary waves form when the wind just begins to blow, and dissipate almost instantly when the driving force is removed.

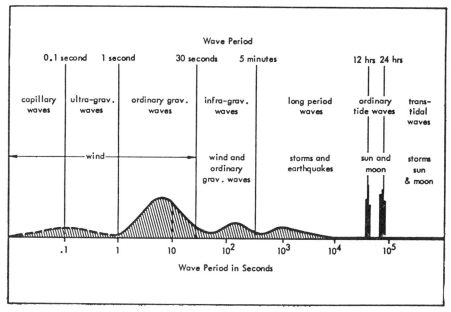

Fig. 1.4. Spectral classification of ocean waves. The relative amplitude is indicated by the curve.

The tides can also be considered as a wave with periods of several hours and wavelengths equal to half the circumference of the earth. Tides are a global response of the oceans to the perturbations of the sun and moon; they oscillate about imaginary nodal (amphidromic) points, as discussed in Chapter 5, with their local effects greatly modified by the topography of the ocean basins.

Other changes in sea surface elevation not falling under the headings of waves or tides are storm surge, associated with steep changes in atmospheric pressure, and seiche or harbor surge, a resonant phenomenon initiated by other sea motions.

The ocean's waters can be characterized by subtle changes in temperature, density, and salinity as functions of space and time. The temperature and salinity, and hence the density, of the open ocean waters remain within relatively close limits worldwide; but locally in coastal waters more drastic changes in these parameters and in the water chemistry may occur, changes that are of particular interest to oceanographers and engineers alike. The ocean's waters can also be thought of as a dilute chemical soup containing all of the natural elements in some form or trace amount. An introduction to the physical and chemical processes of the sea is provided in Chapter 7.

The oceans support a wide diversity of life forms, approximately 160,000 different species or about 16% of all animal species. Biological processes in the sea can be of great importance to ocean engineers, especially as they affect changes in seawater chemistry, noisome growths (fouling), and direct attack on structures by marine organisms (marine borers). The role of bacteria may be most important to the corrosion and fouling history of a structure. Aspects of bio-deterioration are discussed in Chapter 7.

The subject of submarine geology has not been treated in this text. However, the designer of fixed ocean structures must naturally be concerned with the local bottom topography, character, and composition, and the likelihood of earthquakes and/or underwater landslides with the possible consequences of tsunamis or turbidity currents. Recent estimates indicate a worldwide rise in sea level on the order of 4 inches per 100 years, known as the eustatic change in sea level. The ocean engineer, however, must be concerned with more local relative changes in sea level, such as upward warping of the sea floor due to isostatic rebound as the bottom continues to unload from the burden of Pleistocene glaciation, or settlement due to consolidation of bottom sediments.

The topography and character of the sea floor have been shaped by the large-scale geological processes of crustal deformation, including volcanism, orogenesis, and erosion and deposition due to sediment transport by wind and water, and by biological processes such as reef building by coral organisms and the accumulation of the shells and tests of minute organisms to form deep bottom sediments. The major features formed by erosion and deposition are the great continental shelves, which may be nonexistent or may extend outward 200 miles or more from the mainland. Typical depths over the shelves range from 100 to 600 feet (compared to the 12,000-foot mean depth of the oceans), and the bottom is typically sandy or muddy. The continental shelves have come under closer scrutiny within the last few years with the advent of the offshore drill rigs. The continental slopes drop to the sea floor over an average maximum distance of 30 to 50 miles with average slopes varying from 1:6 to 1:40. Beaches are coastal features of prime importance. Beaches tend to be steeper in the winter months owing to the scouring and shoreward movement of sand with increased wave action. Therefore, beaches typically change their profiles on a seasonal basis. Beaches may be classed according to predominant particle size (i.e., cobbles, pebbles, sand, silt, etc.) and steepness. There is generally a direct relationship between surf energy, particle size, and steepness. Beaches tend to be composed of sand-sized particles with typical steepnesses ranging from 1:3 to 1:40. The study of beaches and their protection is an important aspect of coastal civil engineering that is beyond the scope of this text. In general, the high water content, low unit weight, and low shear strength that are characteristic of ocean bottom sediments give rise to difficult foundation conditions.

Selected general references on the oceans and physical oceanography of particular interest to ocean engineers are included at the end of this chapter.[11-13]

1.4 MARINE STRUCTURES

Marine structures vary in sophistication from relatively simple waterfront bulkheads to complex offshore platforms. They may be permanently founded on the seabed, free-floating, or fixed moored (attached to some mooring device that is permanently attached to the seabed or adjacent land). Traditional marine structures can be broadly grouped into the categories of port and harbor structures, which are constructed primarily for the berthing, mooring, and servicing of vessels, and coastal and shore protection structures, which are constructed for the purpose of creating artificial harbors, protecting or reclaiming land space, and protecting or creating recreational beaches. Today we can add another broad category: marine structures that are constructed for the collection of natural resources, such as oil, gas, and minerals, and for the storage and conveyance of such resources. Traditional harbor or waterfront structures include piers, wharves, bulkheads, quay walls, dolphins, dry docks, and navigation structures such as buoys, channel markers, light towers, and locks. Piers and wharves may be of either open

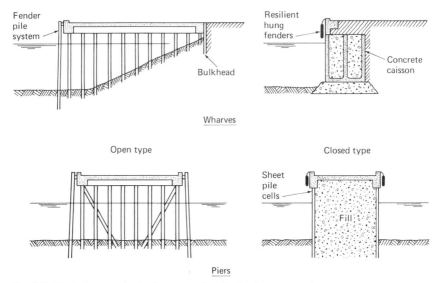

Fig. 1.5. Typical pier and wharf cross sections indicating open and closed type construction.

construction, (i.e., supported by piles or some type of open framework) or closed construction (i.e., constructed of sheet piles, cribwork, caissons, or gravity walls). Figure 1.5 illustrates some pier and wharf structural types, and Fig. 1.6 shows a modern container and general cargo wharf. Piers are generally distinguished from wharves in that piers project out over the water, usually being nearly perpendicular to the shoreline, whereas wharves are usually parallel to and often contiguous with the land. Environmental loadings, in particular wave forces, are dependent upon the structural type and configuration. Dolphins are solitary structures used to breast or moor vessels. Breasting dolphins usually have some form of resilient fendering attached to their exposed faces to aid in absorbing vessel impacts such as are visible on the wharf and attached dolphin shown in Fig. 1.6. Mooring dolphins are designed only to take the lateral and uplift forces of a vessel's mooring lines. Offshore terminals for tankers and bulk cargo vessels, such as the one shown in Fig. 1.7, usually consist of dolphins for breasting and mooring vessels and an intermediate platform for the handling of hoses or conveying equipment, all of which are interconnected by catwalks and are usually attached via a trestle to land. Further discussion of typical water-

Fig. 1.6.   A modern container and general cargo wharf constructed of precast and cast in place concrete founded on precast prestressed concrete cylinder piles. The main wharf deck is 600 feet long and the total berth length over the attached mooring dolphins is 1350 feet. (*Photo Courtesy of Parsons-Brinkerhoff*)

front structures, types, and configurations can be found in the references.[14][16]

Dry docks are used to haul vessels from the water for maintenance and repair. There are four major types of dry docks: (1) The basin or graving dock is essentially a basin with an opening gate called a caisson at one end, through which a vessel is floated; the basin is subsequently dewatered. (2) The floating dry dock is a freely floating hull that is submerged by flooding; a vessel is floated over it, and then it is pumped out in order to raise both dock and ship. (3) The marine railway consists of a cradle riding on inclined ways or rails that transports a vessel up and down the track. (4) The vertical lift dry dock consists of a platform slung between hoisting cables. Although most dry docks are constructed in relatively protected areas, some,

Fig. 1.7. A panoramic view of a contemporary offshore tanker terminal (Sea Island) located in Bantry Bay off the coast of Ireland. The structure has an overall length of 1600 feet and includes four mooring and eight breasting dolphins and provides a 100 foot water depth for mammoth tankers of up to 326,000 tons deadweight. (*Photo Courtesy of P.R.C. Harris, Inc.*)

such as the large floating dry dock shown in Fig. 1.8, must be designed to withstand significant environmental loadings; in this case the floating dry dock had to withstand the rigors of a trans-Pacific ocean tow. This floating dry dock is also an example of a contemporary structure that integrates the disciplines of civil engineering and naval architecture, and which was created in response to changing needs in the marine transportation and energy industries, namely, the advent of the very large crude carriers (VLCC) and the liquid natural gas carriers (LNG) type vessels.

Coastal defense and shore protection structures include: breakwaters, jetties, groins, sea walls, and revetments. These structures can be further defined as: rigid, semi-rigid, or flexible. Rigid structural types include concrete gravity walls and caissons or sheet pile

Fig. 1.8. A contemporary floating drydock of 81,000 long tons lifting capacity with a length overall of 982 feet and breadth of 228 feet. The design of this drydock integrated the disciplines of civil engineering and naval architecture and although it is intended to live out its design life in protected waters, it had to be designed to withstand the rigors of deep ocean towing. *[Photo Courtesy of Port of Portland, Portland, Oregon]*

bulkheads of timber, steel, or concrete. Semi-rigid structures are those that can absorb some degree of wave impact such as steel sheet pile cells or timber cribs. Flexible structures are usually of the rubble mound or rip-rap type construction. The effect of structural type and wave–structure interaction upon design is discussed further in Chapter 3, Section 3.6. Breakwaters are often major civil engineering works which require detailed studies of the local wave climate and the alterations imposed by the introduction of the breakwater itself. In the planning of such a structure model studies are essential. Groins and jetties are built essentially perpendicular to the shoreline in order to prevent the loss of beach material via littoral drift. Seawalls and revetments are built along the shoreline, often at the head of a beach, in order to prevent further shoreline erosion. Details of construction of shore protection structures can be found in the references.[17, 18]

The development of contemporary offshore ocean structures has greatly increased our knowledge of the marine environment. Because of the critical nature of environmental loadings on these structures, funding of research and study efforts were begun. In the earlier decades of offshore oil development such information was largely considered proprietary among the oil companies, but today much of this information has filtered down to the public domain. Offshore oil drilling platforms may be fixed bottom types, usually of tubular steel columns secured to the sea bed with piles or to a large spread mat foundation, or they may be of the concrete gravity platform type, such as the Condeep production platforms developed for the North Sea. Alternatively, offshore drill platforms may be of the compliant type, exhibiting some type of motion in the sea. Compliant type structures are more suitable in deeper water where fixed bottom structures become monstrous in size. Compliant types include: the semi-submersible, with its conventional catenary chain mooring system; the articulated column type, which is attached to the bottom at a single pivot point and utilizes positive buoyancy to remain upright; the guyed tower, which rests on a small foundation base and is held upright by guy lines; and the tension leg platform, which is a positively buoyant platform structure held to the bottom by cables or risers.

Fig. 1.9. A contemporary ocean engineering structure, the design of which incorporated advanced techniques in prestressed concrete technology and careful consideration of environment loadings. This floating LPG storage facility is permanently moored offshore in the Java Sea. The hull is 461 feet long by 136 feet beam with a gross displacement of 66,000 tons. (*Courtesy of Concrete Technology Corp.*)

In addition to offshore drilling platforms, current energy needs call for offshore storage tanks situated entirely below the surface, surface-piercing but fixed to the bottom, or completely free-floating, such as the LPG storage barge shown in Fig. 1.9. The need to construct relatively sophisticated structures in remote locations has brought about the practice of mounting various types of plant and equipment on floating pontoons or barges. Underwater pipelines are used to convey oil and gas ashore from storage tanks and from offshore tanker berths or moorings.

Harnessing the energy of the tides, waves, or thermal differences within the ocean, the possibility of offshore power plants or, floating airports, and the extension of various other activities into the ocean will provide many challenges for the structural engineer, all of which enterprises must begin with a sound understanding of the marine environment.

1.5 THE DESIGN PROBLEM

The structural engineer's problem is to integrate the various environmental factors affecting loadings on and durability of a structure with its other structural, functional, and operational requirements, while weighing safety and reliability against cost and availability restrictions. Environmental loadings may be the major source of loadings and hence the limiting design criteria, or they may be negligible on a given structure. In any case the potential effects of the marine environment deserve close scrutiny in all structures built in and around the sea. The condition of the sea at any time is usually the result of many interacting phenomena. The designer of marine structures must recognize interelationships, and by looking at various phenomena under simplified, idealized conditions will gain some insight into the ocean system at large. Wind, waves, currents, ice, hydrostatic pressure, and impact are the primary environmental sources of loading that a structure must contend with while standing up to various deteriorating agents such as corrosion, attack by marine organisms, wear, and fatigue. The structure also must remain serviceable under the effects of everyday and extreme changes in water levels, possible scouring and deposition of bottom material by waves and currents, freezing and thawing effects, and so on.

The designer of sea structures must also consider the problems of constructing a given design in a harsh environment. In general, the less time required for installation at the site, the greater will be the overall savings in effort and cost. The use of prefabricated elements is warranted for nearshore structures such as piers and wharves, which must be constructed over water, as well as for offshore structures, the installation of which may be possible only within a narrow time frame (window) due to climatic conditions. Problems of marine construction and operations are beyond the scope of this book. However, one must bear in mind that the structural designer must often compromise an otherwise ideal design because of practical problems of installation caused by environmental conditions. Compromise may again be affected by the impracticality of long-term maintenance. In general, it is desirable to minimize the number of members and bracing wherever possible in order to facilitate inspection and maintenance and possibly cleaning and repairs. Accessibility of all portions of the structure is especially important in this regard.

Because of the probabilistic nature of extreme environmental events, it is most desirable that the main structure possess redundancy; in other words, that an alternate load path be provided for all modes in which the structure could otherwise fail catastrophically. Marine structures in general should possess a certain degree of ductility. Breakwaters, for example, have traditionally been designed to some minimum level of acceptable damage criteria with the knowledge that if, say, a 100-year storm occurred at a given time in the structure's life, then certain repairs would be required, but the major part of the structure would remain intact. Waterfront structures have traditionally been designed with a high degree of conservatism, largely because of the unknowns in possible loadings and because of the relatively rapid deterioration rates of structures in the sea. Today, with our better understanding of environmental loadings and statistical approach to determining design criteria, and with high-speed computers to aid analysis, as well as better quality control of materials and protective coatings and systems, we have the ability to produce more economical designs on a rational basis. Caution must still be exercised, however, as our understanding of the marine environment and its effects upon materials and structures is still incomplete, although great progress has been made.

Ultimately the determination of what level of risk is acceptable, the problem of cost versus safety and reliability, is not usually made by the engineer, but it is incumbent upon the engineer to provide the most accurate and plentiful information possible as a basis for such decisions. Further, because such decisions represent a somewhat subjective judgment, the marine structure engineer with his or her knowledge of structural behavior and understanding of environmental conditions must provide the final judgment as to whether the structure is adequate for its purpose. Adequate design can be provided within a reasonable budget by first understanding the forces that will likely act upon the structure during its design life, by understanding the ways in which it may fail or be rendered unserviceable, and by providing, through such understanding, at least the minimal structure that is required. The problem of risk and reliability (the probability that a structure will perform satisfactorily during its lifetime) versus safety and economy has been treated by many authors, notably by Bea[19, 20] and by the earlier works of Marshall,[21] Freudenthal and Gaither,[22] and Borgman.[23] These selected references should prove useful to the reader in relating the following discussion of the marine environment to its ultimate application to structural design. There is much more to be learned in our understanding of the oceans. The design of marine structures will continue to be an important aspect of our advancement into the sea, and an understanding and appreciation of the marine environment will result in more compatible and lasting structures in harmony with the medium surrounding them.

REFERENCES

1. Gaythwaite, J. W. "Structural Design Considerations in the Marine Environment," *Boston Society of Civil Eng'rs. Section, ASCE,* Oct. 1978.
2. Crandall, J. S., Visiting lecturer, unpublished lecture notes on ports and harbors. Harvard University, 1945–1950.
3. Savile, L. H. Presidential address, *Journal of The Institute of Civil Eng'rs.,* London, 1940.
4. Gerstner, F. J. "Theorie der Wellen," *Ann. Physik,* 1809.
5. Airy, G. B. "On Tides and Waves," *Encyclopedia Metropolitania,* 1845.
6. Stokes, G. G. "On the Theory of Oscillatory Waves," *Trans., Cambridge Philosophical Society,* 8, and paper I, 1847.
7. Reeves, H. W. "An Overview of the Offshore Platform Industry," *New England Section, SNAME,* Nov. 1975.
8. Inman, D. L. "Ancient and Modern Harbors: A Repeating Phylogeny," *Proc.* 14th Conf. Coastal Eng'g., ASCE, 1974.
9. Bruun, P. "The History and Philosophy of Coastal Protection," *Proc.* 13th Conf. Coastal Eng'g., ASCE, 1972.
10. Munk, W. H. "Origin and Generation of Waves," *Proc.* 1st Conf. Coastal Eng'g., Council on Wave Research, 1951.
11. Sverdrup, H. V., Johnson, M. W., and Fleming, R. H. *The Oceans,* Prentice-Hall, Englewood Cliffs, N. J., 1942.
12. Defant, A. *Physical Oceanography,* Vols. I and II, Pergamon Press, New York, 1961.
13. Neumann, G., and Pierson, W. J. *Principles of Physical Oceanography,* Prentice-Hall, Englewood Cliffs, N. J., 1966.
14. Quinn, A. D. *Design and Construction of Ports and Marine Structures,* McGraw-Hill, New York, 1972.
15. Bruun, P. *Port Engineering,* Gulf Publishing, Houston, 1976.
16. Cornik, H. *Dock and Harbour Engineering,* Vols. I–IV, Griffin and Co., London, 1962.
17. U. S. Army. *Shore Protection Manual,* Vols. I–III, Coastal Eng'g. Research Center, 1977.
18. Silvester, R. *Coastal Engineering,* Vols. I and II, Elsevier Scientific Publishing Co., New York, 1974.

19. Bea, R. G. "Selection of Environmental Criteria for Offshore Platform Design," *Offshore Technology Conferences (OTC)*, #1839, May 1973.
20. Bea, R. G. "Earthquake and Wave Design Criteria for Offshore Platforms," *Proc. ASCE*, ST-2, Vol. 105, Feb. 1979.
21. Marshall, P. W. "Risk Evaluations for Offshore Structures," *Proc. ASCE*, ST-12, Vol. 95, Dec. 1969.
22. Freudenthal, A. M., and Gaither, W. S. "Design Criteria for Fixed Offshore Structures," *OTC* #1058, Houston, May 1969.
23. Borgman, L. E. "Risk Criteria," *Proc. ASCE*, WW-3, Vol. 89, Aug. 1963.

2
Overwater Wind

Wind loadings and effects certainly are not peculiar to marine structures, but overwater winds tend to be more severe and frequent than wind over land and may occur simultaneously with maximum wave loadings and higher than normal water levels. Marine structures are inherently more susceptible to extreme winds than land structures, as storm systems gather energy from the water and there is no sheltering effect of terrain or vegetation. High winds can entrain water, especially when accompanied by heavy rain, which greatly increases the dynamic pressures. Offshore structures are usually well elevated above the sea bottom, so that wind loads are applied at levels that produce maximum overturning moments. Such structures are also more susceptible to dynamic effects because of their flexibility.

This chapter briefly reviews the nature of storm systems responsible for extreme winds and presents criteria for the selection of design wind speeds. The application of probability and risk criteria is considered. Such criteria are applicable to other probabilistic occurrences of extreme events such as waves and high water levels.

After the basic design velocity has been determined, the wind forces and moments can be evaluated with regard to the vertical distribution, appropriate gust factors, and so forth. Selection of proper drag coefficients is emphasized.

Finally the structure, especially long slender elements, must be checked for possible dynamic effects over a range of wind speeds and with respect to the unsteady nature of the wind. Dynamic response of a structure may occur through various coupling mechanisms such as vortex shedding at a relatively constant velocity and amplification of deflections in periodic gusts wherein resonant frequencies are excited.

2.1   THE NATURE OF THE WIND

Wind, in essence, is a horizontal movement of air in response to differences in air pressure caused primarily by differential heating and cooling. The path of the wind is modified by the rotation of the earth and by topography. The earth is girdled by major surface wind systems such as the trade winds and prevailing westerlies, for example, which are a part of the general circulation of the atmosphere. Major wind systems vary within fairly well-defined latitudinal bands, each of which is subject to short-term disturbances that are responsible for extreme winds. Although the general circulation of the atmosphere and principles of atmospheric stability are beyond the scope of this text, familiarity with basic meteorology is important to those involved in the planning of sea structures. Interested readers are referred to references 1 and 2 or any basic meteorology text for a concise overview of the subject. The structural engineer is necessarily concerned with the extreme winds that may occur at the structure site, although for operational considerations records of mean windspeeds and directions throughout the year may be required. Information on average winds is readily available through the National Oceanic and Atmospheric Administration (NOAA) and the U. S. Naval Oceanographic Office (USNOO) for many sites around the world. Records for extreme winds must necessarily be taken over long periods of time, and often design values are extrapolated from statistical samples.

The general circulation is manifested in its component parts, primarily cyclones and anticyclones (corresponding to low and high pressure centers, respectively), which form an endless procession around the globe. Intense low pressure systems known as extratropical cyclones and tropical cyclones (hurricanes) are responsible for most of the higher wind velocities governing the design of marine structures. The mid-latitudes in particular, where most structures are built, are a zone of atmospheric disturbance. When a temperature difference exists between air masses, a flow of air is set up that attempts to equalize the pressure. Air flows counterclockwise about low pressure centers in the Northern Hemisphere and clockwise in the Southern Hemisphere. The initial boundary between air masses is termed the frontal surface, so that weather trends are typically marked by the alternate passage of warm fronts and cold fronts. Winds associated with intense extratropical storms will usually determine the design winds for sites not exposed to hurricanes.

Tropical cyclones, known as hurricanes in the North Atlantic, typhoons in the western North Pacific, and willie-willies off the east coast of Australia, are large-expanse storm systems, with awesome wind speeds, which occur on a seasonal basis in restricted areas of the world. The western Pacific typhoons are perhaps the most awesome of these storms, as they have the most unrestricted expanse of ocean over which to develop and travel. The occurence and the behavior of tropical storms in various parts of the world have been concisely summarized in reference 3. Detailed information on particular storms can be found through NOAA.[4] Severe hurricanes occur on an average of two per year for the eastern and Gulf coasts of the United States. They are spawned in the tropics and work their way northward or westward as they gather energy from the sea. Hurricanes usually diminish in intensity when they encounter land (see next section for more detail).

Tornadoes and waterspouts are highly localized and extremely intense atmospheric disturbances associated with the highest known wind speeds (up to 200 to 300 MPH or greater). It is not generally considered worthwhile, however, to design structures for such incredible velocities, because of the tremendous forces involved and the low probability of encountering such storms. Major damage to buildings in the direct path of tornadoes appears to result from explosion due to the sudden and severe drop in outside pressure rather than from the dynamic pressure of the wind. The state of the art of tornado-resistant design and tornado behavior has been summarized by Kessler[5] and by Twisdale,[6] and further references may be found in their papers. Such storms are not peculiar to marine structures, however, and are beyond the scope of this text.

Wind systems are affected by topography and are sometimes the direct result of it, such as foehn winds, associated with mountain ridges, or bora winds, which occur in areas where a steep slope separates a cold plateau from a warm plain. Intense and turbulent winds of the foehn type are known as chinook winds in the Rocky Mountains and are associated with rapidly rising temperatures and consequent snow melts. There are anabatic (upslope) and katabatic (downslope) winds and even more local funneling of winds through restricted areas called jets. Monsoon winds are an example of a large-expanse seasonal wind pattern which has a major effect on climate and weather along the northern shores of the Indian Ocean. The wind systems of primary importance to our discussion are those associated with tropical (hurricanes) and extratropical cyclones. Understanding the basic meteorology of these storms is most important, as they affect not only the wind speeds and patterns but also the development of waves and the maximum water levels (storm tide) that may occur at a site.

2.2   DESIGN WIND

General Considerations

The important parameters that define wind criteria for design purposes are: the determination of the basic wind velocity to be used; the spatial variations of the basic wind velocity, in particular the variation with altitude; and the temporal variations or momentary increases in the basic velocity manifested as gusts. Building codes have traditionally specified wind pressures to be used in the design of buildings and other land-based structures. However, the design of marine structures and other wind-sensitive and exposed structures calls for a more accurate assessment of probable loadings than is considered by most building codes. Some pioneering codes for offshore structures such as references 7 and 8, however, call for statistical data to be used as a means of determining the maximum probable winds at a site and list criteria to be considered in determining the appropriate design wind velocities. Where statistical data are available, it is customary to consider the probability of a given wind speed occurring associated with a mean interval of occurrence called the return period. Choice of the appropriate return period depends upon the structure's design life and the probability that the given wind will occur or be exceeded at the site. Once the basic design wind velocity is determined, based on a reference datum, its vertical profile must be considered before forces and moments can be determined. Wind force calculations must also consider appropriate gust factors, which account for peak loads of short duration. Complete

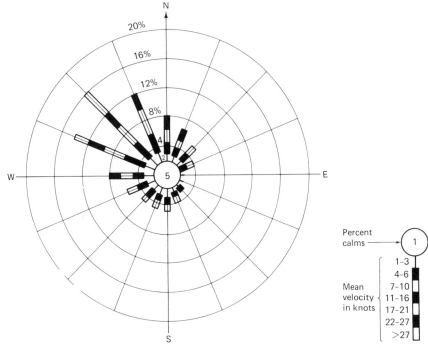

Fig. 2.1. Wind rose. Figure indicates percent frequency of time wind blows from given direction for given month, season, year, etc.

wind data for a site should include information on the average winds, such as monthly mean and maximums, for each month of the year, as well as information on long-term extreme winds. Average winds can be shown on a wind rose, as in Fig. 2.1, which indicates the percent frequency of winds of a given force from a given direction. This information is perhaps more useful for operational and construction considerations than for structural design, but may also be important in the fatigue life history of wind-sensitive structures and in estimating the average wave climate.

Determination of Basic Design Wind Speed

Because the wind exhibits marked temporal variations, an appropriate averaging time interval must be selected so that wind data can be analyzed on a consistent basis. The averaging interval must be long enough that the recorded speed extends over the length of the structure and discounts peak velocities that are of shorter duration than that structure's response time, and be short enough to obtain a realistic maximum velocity from which forces can be computed. In the recording of high wind velocities, the response time of the recording instrument (anemometer) must also be considered. In Great Britain, wind speeds for design have been traditionally averaged over one minute. In the United States, however, it is standard practice to define the basic design wind in terms of the "fastest mile of wind," which corresponds to a gust length of one mile. For example, a 60-knot velocity would have to be sustained over a period of at least one minute to be considered the fastest mile of wind in a record. The relationship between the fastest mile and other time averages will be discussed further subsequently. Further discussion of selection of basic design wind velocities can be found in reference 9. Because the wind velocity increases with height above ground owing to retardation of the lower layers by friction, a standard reference datum must be defined to which wind data can be compared. The anemometer reference height universally used is 30 feet or 10 meters above the surface.

Figures 2.2 and 2.3 show isolines of the fastest mile of wind associated with return periods of 50 and 100 years, respectively, measured at 30 feet above ground for the continental United States. Statistical data used in compiling these charts were taken from airport records and, therefore, are probably low for direct coastal exposures and offshore areas. These charts were presented by H. C. S. Thom[10] based upon his earlier work.[11] The references cited present similar charts for 2-, 10-, and 25-year return periods plus additional data for Hawaii and Puerto Rico. The 50- and 100-year charts only are presented herein, as it is not advisable to design for anything less than these values except perhaps in the case of mobile structures or those erected for temporary use where the potential for loss of life or valuable property is low. Thom has also analyzed the wind climates over the open oceans[12] as they affect wave generation. In areas of the world where statistical records are not available, choice of basic design velocity becomes one of somewhat subjective judgment by the designer based upon what records are available or perhaps consultation of a qualified meteorologist. Sites that lie in the known tracks of tropical storms should be analyzed with respect to whatever data are available on the frequency and severity of such storms and the likelihood of their affecting the site.

Fig. 2.2. Basic wind speed in miles per hour. Annual extreme fastest-mile speed 30 feet aboveground, 50-year mean recurrence interval. (After Thom[10])

Fig. 2.3. Basic wind speed in miles per hour. Annual extreme fastest-mile speed 30 feet aboveground, 100-year mean recurrence interval. (After Thom[10])

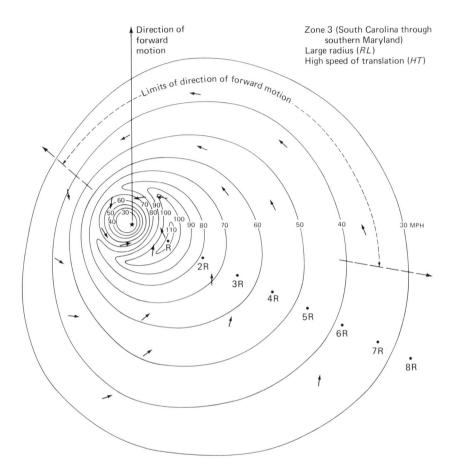

Fig. 2.4. Standard project hurricane, 30-ft. over-water isovel pattern, (After Graham and Nunn[18])

Figure 2.4 is a standard project hurricane (SPH) wind field postulated for the middle Atlantic coast as presented by Graham and Nunn,[13] who similarly present wind fields for various hurricane locations, sizes, and speeds of translation along the U. S. Gulf and Atlantic coasts. Gilman and Myers[14] demonstrate the application of such information to design of structures along the New England coast. The SPH predicated by Graham and Nunn very nearly corresponds to the most severe known hurricane for which reliable data were recorded, off Cape Hatteras, North Carolina, in September 1944. The peculiar needs of the nuclear industry prompted the study of the worst possible storm that could occur in terms of the probable maximum hurricane (PMH).[15] The use of a project hurricane (SPH) or worst probable storm (PMH) for the design of offshore and exposed coastal structures also allows for analysis of the extreme wind and wave conditions that could exist simultaneously at the site.

The important parameters in determining the severity of hurricanes are the hurricane forward speed (V_F), the central pressure reduction from the normal atmospheric pressure (CPI), and the radius to the extreme wind (R), which is the distance from the storm center to the maximum sustained winds. The product $R \times CPI$ is called the energy index and has been used as a measure of a storm's severity in particular regard to its wave-generating potential as discussed in Chapter 3. As winds flow counterclockwise about the storm center, for the hurricane of Fig. 2.4, the maximum velocities are found in the right-hand quadrant in line with the direction of forward motion of the storm. These storms move forward at speeds typically between 5 and 50 knots and gather forward speed with increasing latitude. They may cover areas of as much as 500 miles with hurricane force (60 knots or greater) winds. Reference 13 gives hurricane data (index characteristics) for selected U. S. east coast hurricanes as well as for the SPH. These index characteristics were slightly revised subsequently by the National Weather Service.[16] The highest wind speeds actually recorded in hurricanes are on the order of magnitude of: 115 knots for the 5-minute sustained wind, 130+ knots for the fastest mile of wind, and up to 170 knots for instantaneous gusts. Although it is possible to reasonably predict surface winds by calculation based on pressure gradients and latitude, meteorological conditions must be assumed that are subject to the same vagaries as the assumption of wind velocities. Such calculations are useful, however, in reconstructing the wind fields in storms for which limited measured data are available and are best left to professional meteorologists and oceanographers.

At certain sites, the basic design wind may be modified by local topography causing funneling, updrafts, down drafts, or blanketing effects. Wind is modified by the presence of other structures, and the addition of a structure to a given site may modify the local wind patterns. The directionality of extreme winds may be relevant to a given structure, although for offshore structures it must be assumed that the extreme wind may come from any direction.

Probability and Risk Considerations

Given the present state of the art, the most practical method of obtaining the extreme value of a random event such as high winds is to select a design wind speed associated with a relatively long return period (T_R) such that the probability of encountering such a wind is minimal. The return period corresponds to the average interval of time for which a given event will recur given a sufficiently long period of record. When the return period is known, the probability of encounter (E_P) can be obtained for a given design life (L) of the structure from the following relation:

$$E_P = 1 - \left(1 - \frac{1}{T_R}\right)^L \tag{2.1}$$

Encounter probabilities for typical design lives and return periods are shown in Table 2.1, based on the work of Borgman.[17] It is readily apparent from the table that it is inadequate to design for a condition with a return period equal to the structure life. For example, there is a 64% chance of encountering a 25-year storm in a design life of 25 years. In any given year, there is a 2% chance of encountering a 50-year storm. Although there are other more accurate risk criteria[17] which consider the distribution and severity of damage, they are more cumbersome to apply and may not make significant refinements to the return period–encounter probability criteria.

There are two basic approaches to the prediction of the extreme values of a random variable. The first approach, elaborated upon herein, is the linear extrapolation of relatively short periods of record to cover longer time periods. The second method, which is generally

Table 2.1. Encounter probability vs. design life and return period.

Design life (years)	Return Period (years)						
	5	10	25	50	100	200	500
1	.200	.100	.040	.020	.010	.005	.002
5	.672	.410	.185	.096	.049	.025	.010
10	.893	.651	.335	.183	.096	.049	.020
25	.996	.928	.640	.397	.222	.118	.049
50	.999+	.955	.870	.636	.395	.222	.095

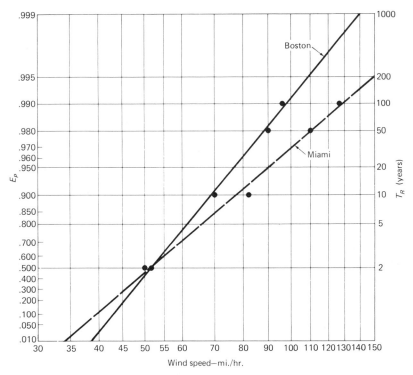

Fig. 2.5. Probability of occurence and return period vs. wind speed for two U. S. Atlantic coast cities. Maximum-value probability paper, Fisher-Tippett type II distribution.

more sophisticated, although perhaps more prone to error, is the prediction of extreme events from statistical models of the actual physical processes involved.[18] The latter is beyond the scope of this text. An example of the former is shown in Fig. 2.5, in which data points from reference 10 are plotted for two U. S. East Coast cities, and fitted to a straight line on a logarithmic extreme value probability paper, as presented in reference 19. In this particular case, sufficient data existed for a 100-year record, which definitely would not be the case for more remote areas. The Det Norske Veritas "Rules for . . . Fixed Offshore Structures"[8] call for the 100-year values to be used for both design wind and wave criteria. Most structures covered by these rules are designed for 10 to 25 years of life. Returning to Fig. 2.5, note that both the locations shown are subject to hurricanes. Winds in excess of 50 miles per hour occur more frequently at

Miami because tropical storms pass on an average of once every year or two, whereas such storms reach Boston only once every 4 to 5 years. Statistical data applied in Fig. 2.4 came from airport records for which wind speeds may be on the order of 10 to 20% less than those for offshore overwater exposures. The probability distribution function used in Fig. 2.4 is known as a Fisher-Tippett type II distribution. Other methods are available such as Beard's method[20] used in hydrologic analysis, Gumbel's method,[21] and the Weibull distribution function,[22] which has particular applicability to wave height distributions.

Gust Factors

Gustiness depends upon the temperature gradient, the mean wind speed, and altitude. Gust factors increase with increasing elevation roughly in accordance with the one-twelfth power of the altitude:

$$\frac{G_z}{G_{30}} = \left(\frac{Z}{30}\right)^{\frac{1}{12}} \tag{2.2}$$

where G_z is the gust speed at elevation Z, and G_{30} is the gust speed at 30 feet above the surface. In applying gust ratios for design, the size and flexibility of the structure must be considered. The maximum gust speed to be applied in force calculations times the gust duration should be at least equal to the extent of the structure in the direction the wind is blowing. Bretschneider[23] summarizes gust factors for various durations for various mean hourly wind speeds. For mean hourly speeds from 20 to 80 knots, the one-minute gust (fastest mile for a 60-knot wind) averages 1.25 times the mean hourly speed, while the 5-second and 0.5-second (instanteous) average gust factors are 1.48 and 1.61, respectively. For short-duration gusts of 0.5 second and 5 seconds, which are within the range of natural periods of most structures, the gust factors (G_{30}/V_1) are 1.3 and 1.2, respectively, for the one-minute average wind speed (V_1). These latter values are consistent with the findings of Gentry,[24] who determined that the mean value of the instantaneous gusts are about 20% higher than the fastest mile, and that the fastest mile was on the order of 20% greater than the 5-minute average sustained velocity between 30 and 110 feet above the surface for several hurricanes.

The possibility of dynamic response of a structure due to wind gusts is discussed in Section 2.4. The possibility of vertical gusts must not be overlooked, particularly in the case of certain vulnerable structures. At heights of 100 feet or more above the surface, vertical gusts can be of the same order of magnitude as the horizontal. Wind speeds in general can be corrected to any duration after the work of Vellozzi and Cohen.[25]

Vertical Distribution

Surface winds extending up to 1000 feet or more in altitude are essentially a turbulent boundary layer flow where the turbulence caused by friction at the air–surface interface is transmitted upward via eddy viscosity. There is some height at which ground friction has a negligible effect on the wind as computed from the pressure gradient, the pressure gradient being the change in atmospheric pressure between two points such as defined by the isobar spacing on a weather map. At this height, the pressure gradient is in equilibrium with the deflecting force owing to the rotation of the earth and the curvature of the wind path. This effect is known as the gradient wind, and the transition from the surface velocity to the gradient velocity is known as the variation with height. The gradient wind is less at the surface than aloft and is dependent upon the sea–air temperature difference. Reference 26 presents some empirical data on the variation with height and makes recommendations on values to be used for design, from which information Table 2.2 has been compiled.

Table 2.2. Stepped values of fastest mile of wind
for coastal areas — variation with height.

Height zone (feet)	Basic wind velocity at 30 feet above ground (MPH)				
30	60	75	90	115	130
0–50	60	75	90	115	130
50–150	85	100	115	140	150
150–400	115	130	145	170	180
400–600	140	160	175	190	195
600–	150	165	180	195	200

Adapted from ASCE.[26]

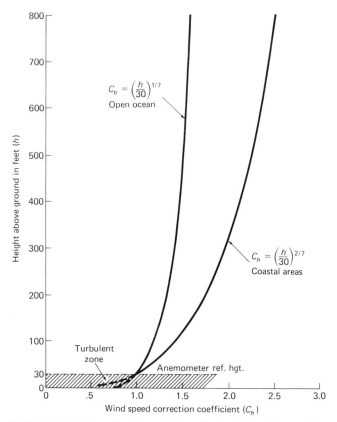

Fig. 2.6. Vertical distribution of wind velocity. C_h is windspeed correction coefficient. (From Gaythwaite[30])

Table 2.2 gives stepped values of wind speeds to be used in design for various basic velocities and heights based on coastal exposures.

Figure 2.6 shows the variation with height in accordance with the wind speed power law, which is used for design of tall structures in which wind forces are significant. It is a semi-empirical law derived from the Prandtl-VonKarman universal velocity distribution law described in almost any fluid mechanics text. The general power law can be written as:

$$V_Z = V_{30} \left(\frac{Z}{30} \right)^n \tag{2.3}$$

where Z is the height above the ground in feet, V_z is the velocity at height Z, and V_{30} is velocity at 30 feet above ground. The exponent

n is usually taken to be between 1/7 and 1/16, depending upon roughness of terrain, or sea condition and distance from land, and the duration of the design wind velocity. The exponent may be around 1/12 to 1/13 for strong gusts, and from 1/7 to 1/8 for sustained winds over the open ocean. For rough coastal areas, it is roughly double the open ocean value, or about 2/7. For further discussion of the application of the power law to design see references 23, 27, and 28. Below the reference height of 30 feet this law is probably not accurate because of great turbulence, and it is recommended that for design the velocity be taken as uniform and equal to that as measured at 30 feet.

2.3   WIND FORCES

When a fixed object is immersed in a stream of flowing fluid, forces arise due to changes in velocity (and hence pressure) around the body. The dynamic pressure exerted by a flowing fluid is given by Bernoulli's equation, which is, in simplified form for our particular case:

$$P = \frac{1}{2} \frac{\gamma}{g} V^2 \qquad (2.4)$$

where P is the dynamic pressure due to the stagnation of the velocity (V), γ is the unit weight of the fluid, and g is the acceleration of gravity. The stagnation pressure P will exist at some point on the body where the flow of fluid is nil and the streamlines separate. Elsewhere on the body, velocities will vary, separating at the edges of the body until they are reunited beyond the turbulent wake. The positive pressure on the upstream face of an object can never exceed the stagnation value; however, as the streamlines accelerate and separate from the edges, a negative (suction) pressure is formed on the downstream face, which can be many times the stagnation pressure. There is an additional force, usually negligible for wind, in the downstream direction due to friction between the fluid and the body. The total force acting in the direction of flow due to pressure difference is called the drag force (F_D), also called form drag or pressure drag. Additionally, a force acting perpendicular to the direction of flow is developed by some shapes or orientations, which is termed the lift force (F_L). Drag and lift forces are usually associated with a moment tending to rotate the body due to an asymmetric pressure distribution about the object's point of fixity. It must be borne in mind that fluid pressures act through the geometric centroid only for symmetrical shapes.

It is customary to apply drag and lift coefficients, which are found experimentally for various geometries, to equation (2.4), in order to obtain the total force. Hence:

$$F_D = C_D \frac{1}{2} \frac{\gamma}{g} V^2 A_P \qquad (2.5)$$

$$F_L = C_L \frac{1}{2} \frac{\gamma}{g} V^2 A_P \qquad (2.6)$$

where C_D and C_L are the coefficients of drag and lift, respectively, and A_P is the projected area exposed to the flow. Drag coefficients for a wide variety of shapes and structures of civil engineering interest are presented in reference 26.

Table 2.3, from reference 26, shows drag and lift coefficients for various structural shapes, which are infinitely long with respect to their other cross-sectional dimension facing the flow. Drag coefficients vary with aspect ratio (λ), surface roughness, and Reynolds number (N_R). Aspect ratio is the ratio of width to length. For square plates normal to the direction of flow ($\lambda = 1.0$), $C_D = 1.12$; for rectangular plates with $\lambda = 5$, $C_D = 1.2$; for $\lambda = 20$, $C_D = 1.42$; and at $\lambda = \infty$, $C_D = 1.98$. Figure 2.7, from reference 29, gives drag coefficients for miscellaneous structures including stacks and towers. Figure 2.8 gives drag and lift coefficients for wires and cables for various angles of attack.

Reynolds number is a dimensionless ratio of inertial to viscous forces, which defines certain flow characteristics.

$$N_R = \frac{VD}{\nu} \qquad (2.7)$$

where D is some characteristic dimension such as diameter, and ν is the kinematic viscosity = .000158 ft^2/sec for "standard" air @ 59°F and 14.70 psia. Figure 2.9 shows the variation of C_D with N_R. Note that for blunt objects with sharp edges such as buildings, deck houses,

Table 2.3. Drag coefficients for structural shapes with $\lambda = \infty$.

Profile and wind direction	C_D	C_L
	2.03	0
	1.96 2.01	0
	2.04	0
	1.81	0
	2.0	0.3
	1.83	2.07
	1.99	-0.09
	1.62	-0.48
	2.01	0
	1.99	-1.19
	2.19	0

From ASCE. [26]

and so on, the variation of C_D with N_R is not significant. For the cylindrical and streamlined objects shown in Fig. 2.9 there is a large decrease in C_D at $N_R = .5 \times 10^6$.

Figure 2.10, from reference 30, is a plot of the drag force equation for various drag coefficients, which can be used for preliminary estimates of wind pressure.

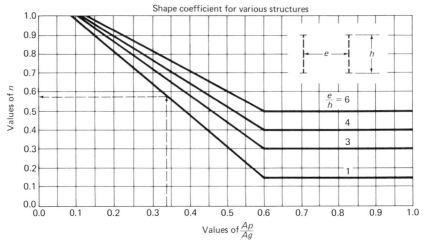

Ap = total projected area of members on one side of the structure.
Ag = total area within the limiting lines for one side of the structure.
F_D = total wind load on the structure. $= P \times C_D \times A_p$
n in the diagram applies to trusses and latticed members except triangular towers.
C_D = shape coefficient

Type of structures	Shape coefficient on projected area s
Double parallel solid girder	1.10
Double parallel trusses and Double parallel latticed members	$1.6\,(1+n)$
Girders and trusses with m parallel members where m is more than 2	$1.5 + (m-2)\,0.5$
Square and rectangular chimneys	1.20
Conical, hemispherical and semielliptical surfaces	0.60
Signboards	1.20
Spheres	0.40
Towers	
Square cross section, wind on face → □	$1.6\,(1+n)$
Square cross section, wind on corner → ◇	$1.92\,(1+n)$
Triangular cross section, wind on face → ▷	$2.28\,(1+n)$
Triangular cross section, wind on corner → ◁	$1.93\,(1+n)$
Use 2/3 of above values for cylindrical members	

Cylindrical surfaces Tanks, riserpipes, chimneys, flagpoles, antennas and similar structures							
$\dfrac{\text{Length}}{\text{Diameter}}$	1	2	3	10	20	40	∞
C_D	0.63	0.69	0.75	0.83	0.92	1.00	1.20

Fig. 2.7. Shape coefficients for miscellaneous structures. (From U. S. Navy[29])

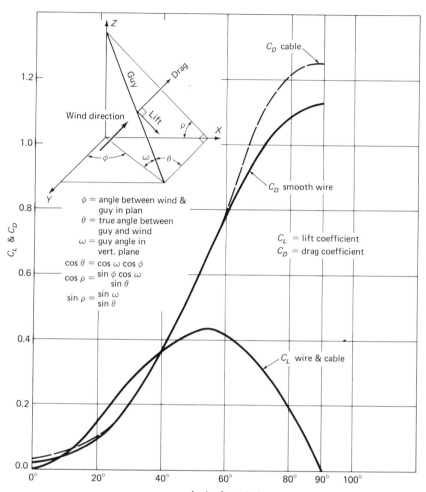

Drag and lift coefficients for wire and cable

Drag & Lift Forces

Drag $= 2.133 \, (c \, d \, v^2) \, C_D \times 10^{-7}$ kips
Lift $= 2.133 \, (c \, d \, v^2) \, C_L \times 10^{-7}$ kips

where c = chord length of cable in feet
d = diameter of cable in inches
v = wind velocity in mph (usually taken as velocity at mid-height of cable)

Note: Lift for leeward cable is positive acting upward.

Fig. 2.8. Wind forces on guy wires and cables. (After U. S. Navy[29])

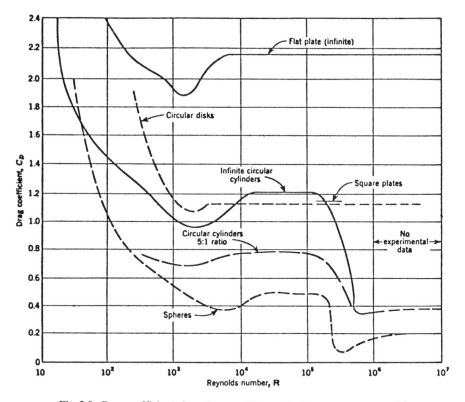

Fig. 2.9. Drag coefficients for spheres and long cylinders. (From A.S.C.E.[26])

The effect of changes in air density with temperature and pressure is frequently neglected. Air density varies linearly with pressure, as a percent of standard pressure, from –10% @ 27 inches Hg (standard pressure is 29.92″ Hg) to +10% @ 33 inches Hg. Air density varies nearly linearly with temperature from +13% @ 0°F (standard temperature is 59°F) to –7% @ 100°F for relatively dry air. These values are affected slightly by moisture content. The unit weight of air at standard temperature and pressure is .07648 lb/cu ft.

In figuring drag forces for complex structures made up of many elements such as trusses and other open frame structures, solidity ratio (Φ) and shielding effects must be considered. The solidity ratio is defined as the ratio of the solid projected area to the total enclosed projected area. For single girders and trusses with small solidity ratio (Φ ⩽ .15) the drag coefficient for the individual member area ap-

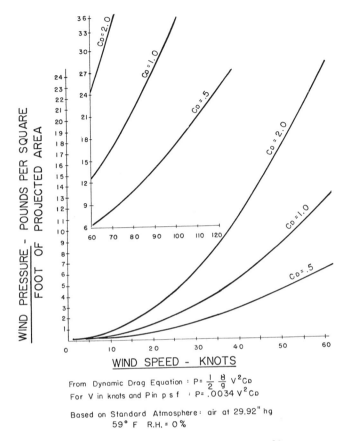

From Dynamic Drag Equation : $P = \dfrac{1}{2}\dfrac{8}{g} V^2 C_D$

For V in knots and P in p s f : $P = .0034\, V^2 C_D$

Based on Standard Atmosphere: air at 29.92" hg
59° F R.H. = 0 %

Fig. 2.10. Wind pressure. (From Gaythwaite[30])

proaches 2.0, while for large solidity ratios ($\Phi \geqslant .90$) the coefficient approaches that for a single large plate with the same aspect ratio.[26] Model test results[31] for common truss types indicate an overall drag coefficient around 1.6 to 1.7. The amount of shielding, or blanketing, of a leeward member by a windward one depends upon the solidity ratio and angle of attack.

Calculation of the total wind force on a structure should of course take into account the vertical distribution of wind speed and temporal variations as discussed in the previous section. These variables are accounted for in most codes or manuals of recommended practice by applying both a coefficient of height factor and a gust factor to equations (2.5) and (2.6). In addition, wind-sensitive structures and those susceptible to dynamic response should be investigated for possible dynamic amplification of loadings as discussed in the following section. In general, objects whose height exceeds approximately four times their least horizontal dimension or whose geometry makes them otherwise wind-sensitive should be investigated in this regard.

In lieu of more specific information, a design wind velocity corresponding to the fastest mile of wind for a 100-year return period measured at 30 feet above sea level, to which the appropriate height and gust corrections have been applied, seems a reasonable guideline for the design of most waterfront, coastal, and offshore structures. Lift forces should always be considered as well as possible dynamic enhancement of forces. Overturning moments may be critical to tall structures and may impose high local foundation loads, such as transmitted by tall cranes on wharves and piers as well as on offshore structures. Thus appropriate safety factors should be applied, especially in consideration of the fact that strong winds also produce high waves and strong currents, which are additive to the wind forces and moments. Further discussion of wind forces on complex structures can be found in references 32 and 33. References 27 and 34 are recent general-reference textbooks on wind loads.

In the design of many large-scale, important structures, and those where the consequences and potential for disaster are great, wind tunnel tests of scale models are warranted. Figure 2.11, from reference 35, is a plot of general model test data for a class EC-2 Liberty ship in a 100-knot wind. The cited reference contains many other model test results, conducted by the U. S. Navy at the David Taylor Model Basin, for different vessel types and wind speeds, along with instructions for applying the data to other similar vessel types. All such test measurements must be scaled from model to prototype in accordance with the laws of mechanical similitude. Since force is a function of projected area, for the test results in Fig. 2.11 the values must be proportioned by the square of the linear ratio. For example, the maximum lateral force at an angle of yaw of 90° on the actual Liberty ship in a 100-knot wind would be:

$$F_{ship} = 78 \text{ lb} \times (83.2)^2 = 539,934 \text{ lb}$$

where 83.2 is the ratio of length of prototype to length of model. For comparison, if the wind area of the actual vessel were known, we

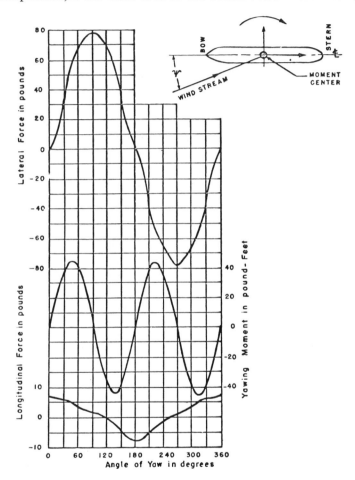

Variation of Lateral Force, Yawing Moment, and Longitudinal Force with Angle of Yaw for a Single 1:83.2-Scale Model of EC-2 Class Liberty Ships; Wind Speed 100 Knots; Moment center at Center of Model

Fig. 2.11. Variation of lateral force, yawing moment, and longitudinal force with angle of yaw for a single 1:83.2–scale model of EC-2 class Liberty ships; wind speed 100 knots: moment center at center of model. (After U. S. Navy[35])

could compute the effective drag coefficient by substituting values into equation (2.5). For example, the lateral wind area of the vessel is, from reference 35, 16,050 square feet; substituting into and rearranging equation (2.5) gives:

$$C_D = \frac{539{,}934 \text{ lb}}{\dfrac{1}{2}\dfrac{.076 \text{ pcf}}{32.2 \text{ fps}^2}(100 \text{ kt} \times 1.69 \text{ fps/kt})^2} \div 16{,}050 \text{ sq ft} = 1.0$$

or the overall drag coefficient for this vessel, side to wind, is 1.0. In this case, the generally accepted practice of using a drag coefficient of 1.3 (see reference 36) or more for moored vessels would be conservative. However, if one considers the wide range of vessel types, unknowns in wind speeds, gust factors, mooring line reactions, and so on, then the value of C_D = 1.3 for the vessel abeam condition is probably a reasonable estimate when model test or prototype data are not available. The curves in Fig. 2.9 could also be used to back-figure the drag coefficient by first calculating the effective wind pressure = 539,934 ÷ 16,050 = 33.6 psf, and intersecting the wind pressure and wind speed on the graph. Reference 37 presents non-dimensional curves yielding wind forces and moments on very large crude carriers (VLCC's) based upon wind tunnel studies. Such information should also be very useful in calculating wind loads on other vessels of similar configuration such as bulk carriers and ore ships.

In lieu of scale model test data, wind forces on moored vessels can be approximated by applying appropriate drag coefficients to the vessels' projected areas as shown in Fig. 2.12. Maximum lateral and longitudinal forces can be reasonably predicted for pier design purposes in this way. However, it is very difficult to estimate yawing moments (rotation of vessel about a vertical axis passing through its center of gravity) because of the complex geometry and lift forces created by a skew wind. Calculation of mooring line loads is beyond the scope of this book – the reader is referred to the cited literature.

Additional considerations in evaluating wind loads include: increase in projected areas of members due to ice accretion (discussed in Chapter 6), and an increase of effective unit weight of air due to

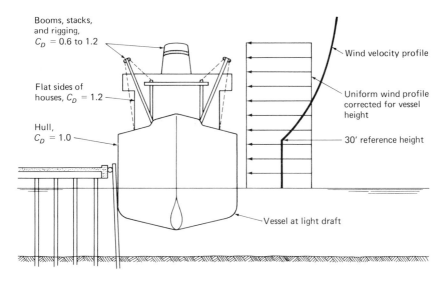

Booms, stacks, and rigging, $C_D = 0.6$ to 1.2

Flat sides of houses, $C_D = 1.2$

Hull, $C_D = 1.0$

Wind velocity profile

Uniform wind profile corrected for vessel height

30' reference height

Vessel at light draft

Notes:

1. Drag coefficients (C_D) shown are representative only and vary with vessel geometry, angle of attack, aspect ratio, surface roughness, etc.
2. Maximum total force occurs with wind abeam; wind at angle creates lift as well as drag forces, which result in yawing moment about vessels C.G.
3. Appropriate gust factor should be applied to above, although short-duration gust forces will be absorbed by vessel response, elasticity of mooring lines, and resiliency of fender system.
4. For pier design it is often assumed that vessel will put to sea or take on ballast to reduce wind exposure when winds exceed storm force (i.e., > 50 knots ±).
5. For rapid estimate of total wind force a factor of $C_D = 1.3$ can be applied to the total projected area of the vessel; this coefficient accounts for both height and gust corrections.

Fig. 2.12. Estimate of wind force on moored vessel.

the entrainment of water from spray, heavy rain, or hail. Little information is available on the latter problem. Reference 23 notes that if air contained only .5% by volume entrained water, the ordinary wind forces would be increased by a factor of five for a 90-knot wind! Recent studies, however, indicate that the effect of wind-blown spray may only double the density of the air as an upper limit[38]. Obviously this is an area for further research, which is especially important to the design of offshore structures. Wind blowing over the tops of large waves may produce a jet of air 30% to 40% greater than the mean wind speed, and researchers[38] have concluded that wind in the gap below the deck of an offshore platform creates low local pressures that tend to create an oscillating vertical load on the platform as waves pass.

2.4   DYNAMIC EFFECTS

General

Unsteady aerodynamic forces may arise on an object in a uniform flow primarily as a result of the shape of the object, or they may arise because of nonuniform turbulent flow resulting in fluctuating loadings. Slender and flexible structures are the most susceptible to dynamic enhancement which depends upon the relation of the structure's natural period in a particular mode of oscillation to the periodicity of the exciting force and the degree of damping (restoring force) in the system. The general equation for the total wind pressure on an object is:

$$P_{\text{total}} = \frac{1}{2} \frac{\gamma}{g} C_D V^2 \pm \frac{\gamma}{g} C_m \frac{A_o}{D} \frac{dV}{dt} \qquad (2.8)$$

where C_m is the mass coefficient, A_o is the area of virtual mass of fluid that moves with the object (= $\pi D^2/4$ for a cylinder), and dV/dt is the time rate of change of wind speed.

The first term of equation (2.8) is the drag force equation previously discussed. The second term is the inertial force, which is generally neglected in static wind load calculations. However, when the wind velocity changes speed and direction very rapidly, as in sharp gusts or tornadoes, the inertial force can be as significant as the static drag force. This is especially true of wind-sensitive structures, which exhibit most or all of the following characteristics: The natural period is not small compared to the gust-acceleration time, the dimensions are such that the gust occurs over a major portion of the structure simultaneously, damping is small, and wind constitutes the major loading. For such structures the dynamic load factor (DLF) may approach its theoretical maximum of 2.0. This means the calculated static force would be doubled for the given wind velocity. If a gust struck in phase with the residual motion from a previous gust, the DLF could exceed 2.0.[39] Design for dynamic loading is beyond the scope of this text. However, some of the important modes of dynamic excitation, shown in Fig. 2.13, are discussed briefly in the following paragraphs.

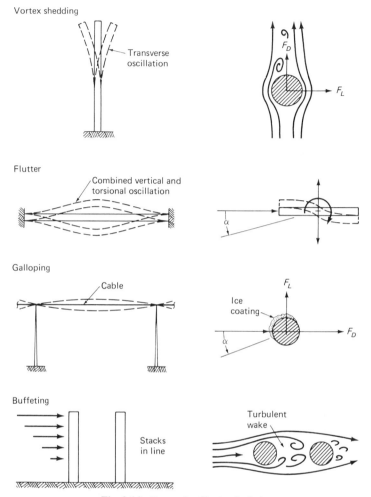

Fig. 2.13. Dynamic effects of wind.

Vortex Shedding

At certain flow speeds turbulent eddies known as vortices may break away from the trailing edge of an object alternately from one side and then the other at a fixed, relatively high frequency. If the period of vortex shedding is near any natural period of vibration of the structure, resonance may result in a self-excited oscillatory motion. This phenomenon is evidenced by transmission lines, tall stacks, signboards, some suspension bridges, and other similar structures. Below some critical speed for the given cross section the rate of vortex shedding is controlled by the wind speed. However, above some critical speed (V_{cr}) the oscillations of the structure, typically transverse to the wind direction, will cause the vortices to be cast away; and thus the frequency of oscillation of the structure and that of the vortex shedding enhance one another. The frequency of vortex shedding is given by the strouhal number (N_s) where:

$$N_S = \frac{fD}{V_{cr}} \qquad (2.9)$$

for which f is the frequency in cycles per second and D is the characteristic dimension of the body. Strouhal numbers for various cross sections are shown in Table 2.4, from reference 26. Strouhal number varies with Reynolds number. For cylindrical bodies with $N_R < 10^5$ it is relatively constant, but approaching $N_R \geqslant 10^6$ it is roughly double that at lower N_R. For flat plates normal to the flow there is little variation with N_R. The possibility of resonance due to vortex shedding can be reduced by changing the cross section or natural period of the structure or by the addition of spoilers.[26]

Flutter

Thin platelike structures parallel to the direction of flow may oscillate in a combined bending and torsion mode when the natural periods of the structure in bending and torsion are nearly equal. The classic example of this was the total destruction of the Tacoma-Narrows suspension bridge in 1940. This structure failed in relatively few cycles of motion and at a moderate wind speed. The wind causes a lift force, which acts eccentrically thus inducing a twisting moment.

Galloping

Cables, such as transmission lines, or other long cylindrical bodies, may exhibit relatively large-amplitude, low-frequency oscillations known as galloping when the system has negative aerodynamic damping. That is, when the lift force shows a negative slope when plotted against

Table 2.4. Strouhal number for a variety of shapes.

Wind	Profile dimensions, in mm	Value of N_s	Wind	Profile dimensions, in mm	Value of N_s
→ ↓	I-section, $t=2.0$, 50, 50	0.120 / 0.137	↓	Channel, $t=1.0$, 12.5, 12.5, 25, 50	0.147
→	H-section, $t=0.5$, 25, 25	0.120	↓	Z-section, $t=1.0$, 12.5, 12.5, 12.5, 50	0.150
↓	H-section, $t=1.0$, 25, 50	0.144	← ↑ ⟋	Angle, $t=1.0$, 50, 50	0.145 / 0.142 / 0.147
↓	T-section, $t=1.5$, 12.5, 50	0.145	← ↑ ⟋	Angle, $t=1.0$, 25, 25	0.131 / 0.134 / 0.137
↓ ↑	Channel, $t=1.0$, 25, 50	0.140 / 0.153	→ ↓	Z-section, $t=1.0$, 25, 25, 25, 25	0.121 / 0.143
↓ ↑	Channel, $t=1.0$, 12.5, 50	0.145 / 0.168	→	Angle, $t=1.0$, 25, 25, 25, 12.5	0.135
→ ↓	Flat bar, $t=1.5$, 50	0.156 / 0.145	→	T-section, $t=1.0$, 50, 100	0.160
	Cylinder $11\,800 < N_R < 19\,100$, 25	0.200	→ ↑	T-section, $t=1.0$, 25, 50	0.114 / 0.145

From ASCE. [26]

the angle of attack (a), the structure will tend to oscillate in the wind. This behavior is typical of ice-coated transmission lines where the coating of ice modifies the cross-sectional shape and thus the aerodynamic lift–drag properties. If the structure is free to oscillate in torsion, an aerodynamic instability will exist when moments are plotted against angle of attacks.

Buffeting

Buffeting action of structures downwind of other, usually similar, structures such as tall stacks in a line is occasionally observed. Buffeting action may occur as a result of resonance with the vortex-shedding frequency of the upstream structure, or may be a result only of the turbulence in the wake of the upstream structure. For many shapes, turbulence may actually reduce the drag force. The reader is referred to the literature for a more detailed discussion of dynamic effects.

The importance of dynamic effects, especially as induced by gustiness, in the design of tall buildings has been recognized in the building code requirements of the American National Standards Institute (ANSI)[40] and the National Building Code of Canada (NBC).[41] Both the ANSI and NBC code requirements attempt to account for the dynamic response of the structure in terms of an equivalent static load, as described by Simiu and Marshall[42] and further by Simiu et al.[43,44] Although the results of these studies may not be directly applicable to most marine structures, the methodology and approach to the problem are of interest.

REFERENCES

1. Donn, W. L. *Meteorology*, McGraw-Hill, New York, 1975.
2. Taylor, G. F. *Elementary Meteorology*, Prentice-Hall, Englewood Cliffs, N. J., 1963.
3. "Special Hurricane Report," *Ocean Industry Magazine*, Houston, Oct. 1973 and Nov. 1973.
4. *Mariners Weather Log*, NOAA, Washington, D.C., published bi-monthly.
5. Kessler, E. "Tornadoes: State of Knowledge," *Proc. ASCE*, ST-2, Vol. 104, Feb. 1978.

6. Twisdale, L. A. "Tornadoe Data Characterization and Windspeed Risk," *Proc. ASCE*, St-10, Vol. 104, Oct. 1978.

7. *Planning, Designing and Constructing Fixed Offshore Platforms*, American Petroleum Institute, RP-2A, Dallas, 1979.

8. "Rules for the Design, Construction and Inspection of Fixed Offshore Structures," Det Norske Veritas, Oslo, 1974.

9. Davenport, A. G. "Rationale for Determining Design Wind Velocities," *Proc. ASCE*, ST-5, Vol. 86, May 1960.

10. Thom, H. C. S. "New Distributions of Extreme Winds in the U. S.," *Proc. ASCE*, St-7, Vol. 94, July 1968.

11. Ibid. "Distributions of Extreme Winds in the U. S.," *Proc. ASCE*, St-4, Vol. 86, April 1960.

12. Ibid. "Distributions of Extreme Winds over Oceans," *Proc. ASCE*, WW-1, Vol. 99, Feb. 1973.

13. Graham, H. E. and Nunn, D. E. *Meteorological Considerations Pertinent to Standard Project Hurricanes, Atlantic and Gulf Coasts of the U. S.*, U. S. Dept. of Comm., N.H.R.P. #33, Washington, D.C., 1959.

14. Gilman, C. S., and Myers, V. A. "Hurricane Winds for Design Along the New England Coast," *Proc. ASCE*, WW-2, Vol. 87, May 1961.

15. U. S. Weather Bureau. *Interim Report — Meteorological Characteristics of the Probably Maximum Hurricane, Atlantic and Gulf Coasts of the U. S.*, U. S. Dept. of Comm., HUR-7-97, 1968.

16. National Weather Service. "Revised Standard Project Hurricane Criteria for the Atlantic and Gulf Coasts," NWS memo HUR 7-120, June 1972.

17. Borgman, L. "Risk Criteria," *Proc. ASCE*, WW-3, Vol. 89, Aug. 1963.

18. Ibid. "Extremal Statistics in Ocean Engineering," *Proc. ASCE*, Civil Eng'g. in the Oceans III, 1975.

19. Thom, H. C. S. "Frequency of Maximum Wind Speeds," *Proc. ASCE*, ST-4, Vol. 80, 1954.

20. Beard, L. "Statistical Analysis in Hydrology," *Trans. ASCE*, Vol. 108, 1943.

21. Gumbel, E. J. *Statistics of Extremes*, Columbia University Press, New York, 1958.

22. Weibull, W. "A Statistical Distribution Function of Wide Applicability," *ASME*, Annual Mfg. Applied Mechanics Div., 1951.

23. Bretschneider, C. L. "Overwater Wind and Wind Forces," in *Handbook of Ocean and Underwater Engineering*, McGraw-Hill, New York, 1969.

24. Gentry, R. "Wind Velocities During Hurricanes," *Trans. ASCE*, Vol. 120, No. 2731, 1955.

25. Vellozzi, J., and Cohen, E. "Gust Response Factors," *Proc. ASCE*, ST-6, Vol. 94, June 1968.

26. Task Committee on Wind Forces. "Wind Forces on Structures," *Trans. ASCE*, Vol. 126, No. 3269, 1961.

27. Simiu, E., and Scanlan, R. H. *Wind Effects on Structures: An Introduction to Wind Engineering*, John Wiley & Sons, New York, 1978.

28. Sherlock, R. H. "Variation of Wind Velocity and Gusts with Height," *Trans. ASCE*, Vol. 118, No. 2553, 1953.
29. U. S. Navy. *Design Manual: Structural Engineering*, NAVFAC, DM-2, Washington, D. C., Oct. 1970.
30. Gaythwaite, J. W. "Structural Design Considerations in the Marine Environment," *Boston Society of Civil Eng'rs. Section, ASCE*, Oct. 1978.
31. Biggs, J. M. "Wind Loads on Truss Bridges," *Trans. ASCE*, Vol. 119, 1954.
32. Pagon, W. "Wind Forces on Structures: Plate Girders and Trusses," *Proc. ASCE*, ST-4, Vol. 84, July 1958.
33. Cohen, E. and Perrin, H. "Design of Multi-Level Guyed Towers; Wind Loading," *Proc. ASCE*, ST-5, Vol. 83, Sept. 1957.
34. Houghton, E. and Carruthers, N. *Wind Loads on Buildings and Structures*, Halsted Press, London, 1976.
35. U. S. Navy. *Design Manual: Harbor and Coastal Facilities*, NAVFAC, DM-26, Washington, D. C., 1968.
36. Quinn, A. D. *Design and Construction of Ports and Marine Structures*, McGraw-Hill, New York, 1972.
37. Oil Companies International Marine Forum. *Wind and Current Loads on VLCC's*, London, 1976.
38. "Oscillating Wind Load and Sea Spray Forces on Offshore Structures," *Research in Ocean Engineering*, bulletin of the Marine Advisory Service of MIT Sea Grant Program, Spring 1979.
39. Norris, C. et al. *Structural Design for Dynamic Loads*, McGraw-Hill, New York, 1959.
40. *American National Standard A58.1 — Building Code Requirements for Minimum Design Loads in Buildings and Other Structures*, ANSI, New York, 1972.
41. *National Building Code of Canada*, National Research Council, Ottawa, 1970, and Supplement #4, 1975.
42. Simiu, E., and Marshall, R. D. "Wind Loading and Modern Building Codes," *ASCE* Nat. Struct. Eng'g. Mfg., Cincinnati, April, 1974.
43. Simiu, E. "Equivalent Static Wind Loads for Tall Building Design," *Proc. ASCE*, ST-4, Vol. 102, April 1976.
44. Simiu, E., Marshall, R. D., and Haber, S. "Estimation of Alongwind Building Response," *Proc. ASCE*, ST-7, Vol. 103, July 1977.

3
Effects of Waves

Waves are periodic undulations of the sea's surface, the complexity of which is most challenging to those working in the oceans. They impose highly variable and fatigue-type loadings on offshore and ex-posed coastal structures, and they may adversely affect coastlines and harbor facilities and induce violent motions in moored ships and floating structures. Waves cause scouring and erosion and may over-top coastal protection structures or exert impact forces and buoyant uplift on structures not situated above the maximum crest elevations. The traditional aspects of coastal engineering (i.e., shoreline changes and littoral transport) are beyond the scope of this text, but the designer of coastal structures must be aware of such problems, with regard to possiblc effects on the structure and the structure's effect on the local environment. Thus appropriate literature is cited wherever possible.

The primary cause of waves is the wind, especially those waves that cause direct structural loadings and violent motions. Therefore, most of this chapter deals with the generation, form, and theory of wind waves. Impulsively generated waves such as those caused by underwater earthquakes, landslides, or explosions behave similarly to those generated by wind, but are generally of much longer period; these waves are discussed in Section 5.5. Other special cases of long-period wave motions such as harbor oscillations, storm surge, and the tides themselves are also covered in Chapter 5.

The designer of marine structures must have a good understanding of the basics of wave generation, characteristics, and behavior before he/she can begin to evaluate forces and other potential effects. At offshore locations and some exposed sites determination of wave climate and possible maximum waves is best left to experienced oceanographers or specialists in this field; but the designer must be able to relate to the work of other disciplines and will ultimately be responsible for the selection of a design wave or waves and/or wave spectrum from which structural loadings are derived. The practicing engineer should also be capable of making a preliminary assessment of potential waves and wave effects at a given site and of determining when a more detailed scientific study is warranted. This also requires a knowledge of how deep-water wave properties are modified by topography and interaction with other sea motions.

This chapter, therefore, first looks at the wave phenomena descriptively, examining wave form and characteristics; next, the principles of wave generation and forecasting are discussed; then some of the mathematical theories that can be applied to the determination of forces and moments are briefly described, the types of forces produced by waves are reviewed and evaluated, and finally other ways in which waves and structures interact are considered.

3.1 THE SEA STATE

It must always be borne in mind that the sea surface at any given time is in a very complex condition defying precise description. Successive wave heights and lengths are not generally equal as they are necessarily assumed to be in theory. Not only does the sea vary from wave to wave, but a given wave crest is not continuous along its own length, a phenomenon known as short crestedness. Waves of varying heights and lengths being driven by the wind, called sea, are usually present with longer, flatter, but more regular swell, which is the term for waves that are decaying after leaving some remote generating area. Wave records taken at a point also do not account for the direction of travel of waves passing that point. All of this variability gives a seemingly hopelessly random situation. However, although one cannot give a precise description of the sea surface at a given time, by observing an area of the sea surface over a period of time one can usually observe a certain characteristic pattern.

Referring to Fig. 3.1, the wave height (H) is defined as the vertical distance between the crest and successive trough, the length (L) is the distance between successive crests, the period (T) is the time required for the passage of successive crests, and the celerity (C) is the velocity of propagation of the wave form, equal to L/T. (The quantity

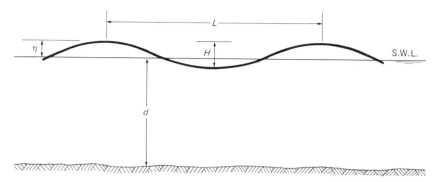

Fig. 3.1. Wave definition sketch.

d is the water depth, and η is the wave crest elevation.) These values are well defined for a simple unidirectional, monochromatic wave. In a typical sea surface record, however, as shown in Fig. 3.2, it is readily apparent that it becomes difficult to define any uniform height or period; thus an apparent height (H_A) or period (T_A) is selected as shown. In this figure the apparent heights and periods have been defined using the zero-upcrossing method, in which heights are defined between troughs and the successive peak height above the still water level (S.W.L.). Another method of analyzing irregular wave trains called the trough–crest method tends to filter out the shorter wavelengths and heights and, therefore, gives slightly different results.

The zero-upcrossing method is generally used in the analysis of wave records taken at a point, but it must be emphasized that none of these methods accounts for the directionality of the irregular wave

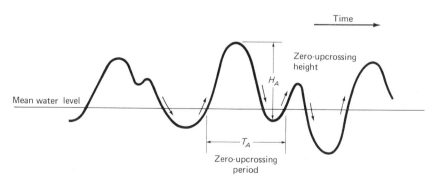

Fig. 3.2. Typical sea surface record.

Two sine waves of different length and amplitude

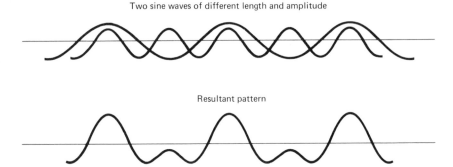

Resultant pattern

Fig. 3.3. Superposition of sine waves.

trains. Oceanographers attempt to study the sea surface by representing the irregular surface pattern with a series of small-amplitude sine waves of varying amplitude and phase. A very irregular pattern can be obtained by the superposition of only a few sine curves; see Fig. 3.3. The phase of the sine curves is random and therefore suggests the use of probability theory. It has been observed that successive points at equal intervals of time in an irregular record follow the normal, or Gaussian, distribution of statistics. There is, therefore, a spectrum of waves describing the sea state as shown in Fig. 3.4, where for any particular period there exist various heights. It is generally more convenient to use the frequency spectrum, which is directly interchangeable with the period spectrum. The wave period is related to the frequency (ω) by: $T = 2\pi/\omega$. These relations and example sea spectra will be discussed further in the next section.

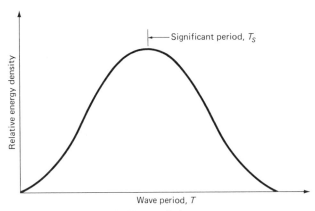

Fig. 3.4. Wave period spectrum.

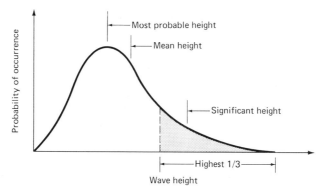

Fig. 3.5. Wave height probability distribution.

It is convenient now to introduce the wave height probability distribution and some fundamental concepts and definitions that are important to an understanding of the rest of this chapter. For a given location and sea state the probability of occurrence of a given wave height roughly follows the Rayleigh distribution, as shown schematically in Fig. 3.5. From this figure a mean height, most probable height, highest of a given percentile, etc., can be obtained. Oceanographers define the average of the highest one-third as the significant wave height (H_s), which corresponds reasonably well with values reported from visual observations, thus giving a convenient yardstick for reporting wave observations. Other statistical heights are related to the significant height as shown in Table 3.1.

The celerity of waves in deep water is governed almost solely by the wavelength; but as the same wave enters successively shallower

Table 3.1. Relation of wave height parameters to significant wave height.

Significant height	1.00
Average height.	0.64
Average of highest 10%.	1.29
Average of highest 1%	1.68
Highest.	1.87
Highest of simple sine waves having same energy content as wave train	0.80

Tabulation based upon cumulative Rayliegh probability distribution.

water and begins to "feel bottom," the water depth controls the celerity. A somewhat arbitrary but convenient definition of deep and shallow water relative to wavelength is as follows: When the water depth (d) is greater than one-half the wavelength ($d > L/2$), the wave behaves as a deep water wave, and the celerity is governed by the wavelength; when $d < L/2$, the wave is considered a shallow water wave and the celerity is governed by the water depth. The general expression for wave celerity, readily derived from Airy's small amplitude wave theory,[1] which is discussed in Section 3.4, is, neglecting surface tension:

$$C^2 = \frac{gL}{2\pi} \tanh \frac{2\pi d}{L} \qquad (3.1)$$

This equation reduces approximately to:

$$C = \sqrt{\frac{gL}{2\pi}}$$

When $d > L/2$ (deep water), for L in feet and $g = 32.2$ feet per second squared, and C in feet per second, with T in seconds $= L/C$, this equation gives:

$$C = 5.12\, T \qquad (3.2)$$

and

$$L = 5.12\, T^2 \qquad (3.3)$$

For shallow water ($d < L/2$) this expression reduces to:

$$C = \sqrt{gd} \qquad (3.4)$$

These relations are plotted for convenience in Fig. 3.6, from reference 2, and their derivation is discussed in Chapter 3.4.

Waves generally begin to break when their steepness (H/L) exceeds 1/10, approaching the theoretical limit 1/7, or when they enter water of depth less than approximately $1.28H$. Shoaling waves shorten in length and build in height, while their period remains almost constant. The limiting depth of breaking waves is most important, as the maximum wave height for a given site may be governed by water depth; in that case such a wave would be relatively steep and perhaps breaking on or near the structure. Wave steepness normally observed at sea

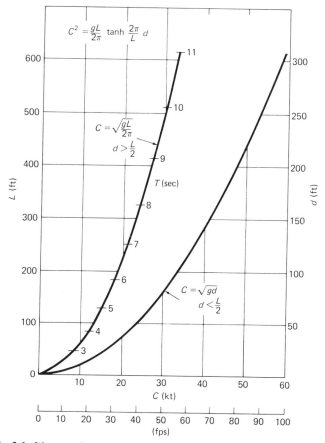

Fig. 3.6. Wavespeed vs. wavelength and water depth. (From reference 2.)

is usually within the range of H/L between 1/17 and 1/33, the steeper waves generally occurring at higher wind speeds and during earlier stages of wave development, and over shallower water. For many years naval architects used a standard $L/20$ wave with L equal to the vessel's waterline length, in determining the longitudinal bending moment due to the sea. Today, however, with high-speed computers in routine use, sea spectra can be applied to analyze the worst bending cases from a simulated random sea. The $L/20$ wave, however, remains a reasonable assumption for the preliminary investigation of bending moment in floating structures. For inland lakes, typically shallow water, wave steepnesses typically range from 1/9 to 1/15. The designer

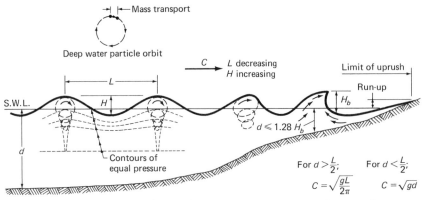

Fig. 3.7. Summary sketch of wave characteristics. (wave propagating toward coast).

of harbor works or other structures in locations where the waves are generated over relatively shallow water will have to contend with relatively steep waves.

Waves normally have more of their height above the S.W.L. than below it in order that the volume of water in the crests equal that removed from the trough. The wave crest elevation (η_c) above the S.W.L. is usually between .55H and .75H, which should be considered when figuring bottom clearance of fixed platform structures.

The movement of a wave corresponds to the transmission of energy. As previously mentioned, it is the wave form that is moving and not the water surface itself. However, if a single wave crest is observed over time, it will seem to eventually disappear as it transmits its energy to another crest. The group velocity of a wave train corresponds to the average rate of transmission of energy and is equal to one-half the wave celerity for a group of regular waves traveling in deep water. As the wave form passes a point on the surface, an individual water particle undergoes a circular orbital motion as shown in Fig. 3.7. The water particle orbits decrease exponentially with depth and become more elliptical, approaching a nearly to-and-fro motion at the sea bed, as the wave propagates into progressively shallower water. The orbit of an individual water particle does not close exactly; there is, in fact, a slight displacement in the direction of the wave advance known as mass transport. This is illustrated in Fig. 3.7, which also summarizes many aspects of wave motion just discussed. Further aspects of wave motion will be discussed in Section 3.5 in connection with wave theories. With some basic principles and definitions having been given, we now turn to a discussion of wave generation and forecasting principles.

3.2 GENERATION, STATISTICS, AND FORECASTING OF WIND WAVES

Generation of Wind Waves

When two adjacent layers of fluid move at different velocities, a frictional shearing stress is developed between them. Thus, a flow instability exists at the air–water interface as the wind begins to blow. According to Phillips,[3] turbulent pressure fluctuations and variations in wind velocity cause an initial disturbance of the sea surface, thus precipitating wave growth. The initial movement of the sea surface is perpendicular to that of the wind, but it quickly aligns itself approximately with the wind direction or with some oblique angle to the wind. A light air movement of only two knots or less will cause small ripples (sailors call them cat's paws) on the sea surface, which disappear immediately when the generating wind stops. These ripples are associated with capillary force and are in conflict with surface tension. As the wind speed increases, and if it is of any duration, larger "gravity waves" begin to develop as the wave height builds up, and a more complex pressure distribution forms at the surface. Normal stresses as well as tangential stresses now act on the surface profile to further wave development. When air flow separates from the wave profile at the crest, form drag becomes significant.

The exact way this growth begins is still not completely understood. There are, however, many semi-empirical relationships that describe the growth of wind waves reasonably well. The ultimate state of wave growth depends primarily on three parameters basic to wave forecasting: the fetch (F) or the distance over which the wind blows, the wind velocity (V), and the duration (t) of time for which the wind blows. Thus, for a given steady wind speed, the development of waves may be limited by the fetch, or the duration. If, however, the wind blows over a sufficient distance for a sufficient length of time, a more-or-less steady state condition, where the average wave heights do not change, will occur. This condition is called a fully developed sea (FDS) and is basic to wave forecasting. The FDS is the wind-speed-limited case. Table 3.2, reproduced from reference

Table 3.2. Wind and sea scale for fully arisen sea.

Sea state	Description	Beaufort wind force	Description	Range, knots	Wind velocity, knots	Wave height, ft — Average	Wave height, ft — Significant	Wave height, ft — Average 1/10 highest	Significant range of periods, sec	$\bar{T}$max, period of maximum energy of spectrum	$\bar{T}$ average period	$\bar{l}$ average wavelength	Minimum fetch, nmi	Minimum duration, hr
0	Sea like a mirror.	0	Calm	Less than 1	0	0	0	0						
	Ripples with the appearance of scales are formed, but without foam crests.	1	Light airs	1–3	2	0.05	0.08	0.10	Up to 1.2 sec	0.7	0.5	10 in.	5	18 min
1	Small wavelets, still short but more pronounced; crests have a glassy appearance, but do not break.	2	Light breeze	4–6	5	0.18	0.29	0.37	0.4–2.8	2.0	1.4	6.7 ft	8	39 min
	Large wavelets, crests begin to break. Foam of glassy appearance. Perhaps scattered white horses.	3	Gentle breeze	7–10	8.5	0.6	1.0	1.2	0.8–5.0	3.4	2.4	20	9.8	1.7 hr
					10	0.88	1.4	1.8	1.0–6.0	4	2.9	27	10	2.4
2	Small waves, becoming larger; fairly frequent white horses.	4	Moderate breeze	11–16	12	1.4	2.2	2.8	1.0–7.0	4.8	3.4	40	18	3.8
					13.5	1.8	2.9	3.7	1.4–7.6	5.4	3.9	52	24	4.8
3					14	2.0	3.3	4.2	1.5–7.8	5.6	4.0	59	28	5.2
					16	2.9	4.6	5.8	2.0–8.8	6.5	4.6	71	40	6.6
4	Moderate waves, taking a more pronounced long form; many white horses are formed (chance of some spray).	5	Fresh breeze	17–21	18	3.8	6.0	7.8	2.5–10.0	7.2	5.1	90	55	8.3
					19	4.3	6.9	8.7	2.8–10.6	7.7	5.4	99	65	9.2
					20	5.0	8.0	10	3.0–11.1	8.1	5.7	111	75	10
5	Large waves begin to form; the white foam crests are more extensive everywhere (probably some spray).	6	Strong breeze	22–27	22	6.4	10	13	3.4–12.2	8.9	6.3	134	100	12
					24	7.9	12	16	3.7–13.5	9.7	6.8	160	130	14
					24.5	8.2	13	17	3.8–13.6	9.9	7.0	164	140	15
					26	9.6	15	20	4.0–14.5	10.5	7.4	188	180	17
6	Sea heaps up and white foam from breaking waves begins to be blown in streaks along the direction of the wind (spindrift begins to be seen).	7	Moderate gale	28–33	28	11	18	23	4.5–15.5	11.3	7.9	212	230	20
					30	14	22	28	4.7–16.7	12.1	8.6	250	280	23
					30.5	14	23	29	4.8–17.0	12.4	8.7	258	290	24
					32	16	26	33	5.0–17.5	12.9	9.1	285	340	27

7	Moderately high waves of greater length; edges of crests break into spindrift. The foam is blown in well-marked streaks along the direction of the wind. Spray affects visibility.		34	19	30	38	5.5–18.5	13.6	9.7	322	420	30
8	Fresh gale	34–40	36	21	35	44	5.8–19.7	10.3	10.3	363	500	34
			37	23	37	46.7	6–20.5	14.9	10.5	376	530	37
			38	25	40	50	6.2–20.8	15.4	10.7	392	600	38
			40	28	45	58	6.5–21.7	16.1	11.4	444	710	42
9	High waves. Dense streaks of foam along the direction of the wind. Sea begins to roll. Visibility affected.	41–47	42	31	50	64	7–23	17.0	12.0	492	830	47
			44	36	58	73	7–24.2	17.7	12.5	534	960	52
			46	40	64	81	7–25	18.6	13.1	590	1110	57
10	Very high waves with long overhanging crests. The resulting foam is in great patches and is blown in dense white streaks along the direction of the wind. On the whole, the surface of the sea takes a white appearance. The rolling of the sea becomes heavy and shocklike. Visibility is affected.	48–55	48	44	71	90	7.5–26	19.4	13.8	650	1250	63
	Whole gale*		50	49	78	99	7.5–27	20.2	14.3	700	1420	69
			51.5	52	83	106	8–28.2	20.8	14.7	736	1560	73
			52	54	87	110	8–28.5	21.0	14.8	750	1610	75
			54	59	95	121	8–29.5	21.8	15.4	810	1800	81
11	Exceptionally high waves (small and medium-sized ships might for a long time be lost to view behind the waves). The sea is completely covered with long white patches of foam lying along the direction of the wind. Everywhere the edges of the wave crests are blown into froth. Visibility affected.	56–63	56	64	103	130	8.5–31	22.6	16.3	910	2100	88
	Storm*		59.5	73	116	148	10–32	24	17.0	985	2500	101
12	Air filled with foam and spray. Sea completely white with driving spray; visibility very seriously affected.	64–71	>64	>80‡	>128‡	>164‡	10–(35)	(26)	(18)	?	?	?
	Hurricane*											

* For hurricane winds (and often whole gale and storm winds) required durations and fetches are rarely attained. Seas are therefore not fully arisen.

‡ For such high winds, the seas are confused. The wave crests blow off, and the water and the air mix.

SOURCE: W.A. McEwen and A.H. Lewis, "Encyclopedia of Nautical Knowledge," p. 483, Cornell Maritime Press, Cambridge, Md., 1953. "Manual of Seamanship," pp. 717-718, vol. II, Admiralty, London, H.M. Stationery Office, 1952. Pierson, Neumann, James, "Practical Methods for Observing and Forecasting Ocean Waves," New York University College of Engineering, 1953.

Reproduced from *Handbook of Ocean and Underwater Engineering.*[4]

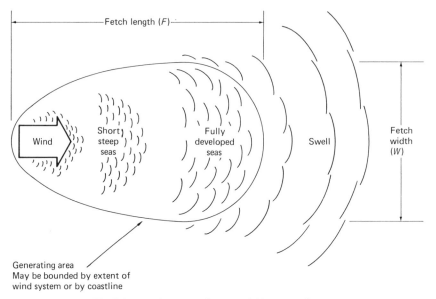

Fig. 3.8. Development of waves within generating area.

4, summarizes sea conditions for fully arisen seas corresponding to various wind speeds. This table utilizes the Beaufort wind and sea scale used by mariners for estimating wind speed from the sea state.

The concept of a significant wave (H_s or $H_{1/3}$), which has already been introduced, is basic to forecasting and observation of ocean waves. If the significant wave height is known, then other statistical parameters can be obtained, as given in Table 3.1, based on statistical probability distributions to be discussed.

Figure 3.8 shows a plan view of a wave generating area, with smaller, generally steeper and short-crested waves being found in the early stages of generation, and longer, more regular waves being found farther downwind. Waves within the generating area still being driven by the wind are called sea, and those leaving the generating area or continuing on after the generating wind has ceased are termed swell. Swell, which are free waves (sea being forced waves), are attenuated by friction and spreading but may travel great distances before being completely diminished by air resistance, turbulence, dispersion, and lateral spreading.

The modulus of decay (τ_p) is the time required for the wave height to be reduced in the ratio of $e:1$ and is given by:

$$\tau_\nu = \frac{L^2}{8\pi^2\nu} \tag{3.5}$$

where ν is the kinematic viscosity. Bretschneider[5] presented the wave decay curves shown in Fig. 3.9, which can be used to determine the increase in significant wave period and decrease in significant wave height at the end of a decay distance as a function of the significant height and period at the end of the fetch.

Wave age is the ratio of wave celerity (C) to the wind speed (V) and is a significant factor in wind wave development. Larger waves may travel faster than the generating wind. When C/V reaches a value of about 1.37, a fully developed sea is attained, and it becomes difficult for the wind to impart further energy to the waves. Within an FDS wave ages may range from 0.1 to 2.0.[6] The average maximum wave speed is slightly greater than the wind speed for wind speeds less than 25 knots and slightly less than the wind speed at higher wind speeds. Also, there is some critical wind speed required, approaching 12.5 knots, for true gravity wave generation. There is no well-defined relationship between wave age (C/V) and wave steepness (H/L). However, waves in early stages of development tend to be steeper than older waves. This relationship is important to theoretical forecasting relations. The relation between wave age and steepness is illustrated in Fig. 3.10 from observed values as reported by the U.S. Naval Oceanographic Office.[6]

The width (W) of the generating area has a noticeable effect on wave development for narrow fetches where $W < 2F$; Saville[7] found effective wave heights of approximately 23%, 65%, 88%, 98%, and 100% for $W/F = 0.1, 0.5, 1.0, 1.5,$ and 2.0, respectively. This may be a factor in narrow harbors and channels. However, in the open ocean W/F is usually greater than unity, and fetch width is seldom an important factor.

Empirical Relations

An early attempt (Stevenson, 1886) to estimate the maximum wave height at a site considered only the fetch length and is approximate for fetches greater than 50 miles:

$$H = 1.5 \sqrt{F} \tag{3.6}$$

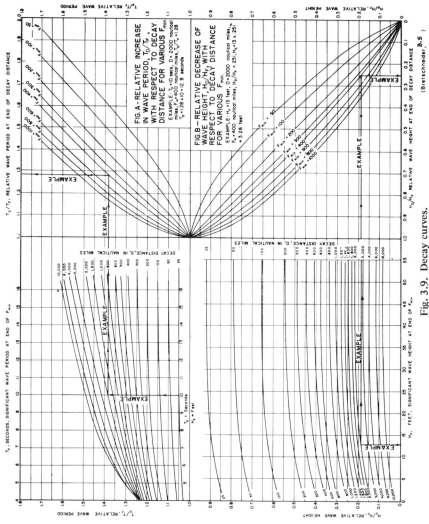

Fig. 3.9. Decay curves.

(Bretschneider, 3.5)

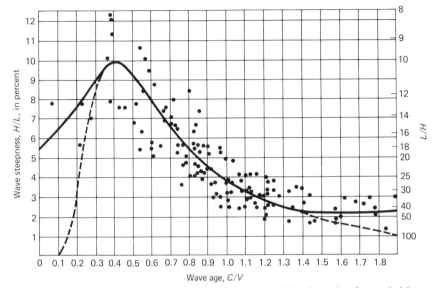

Fig. 3.10. Relation between wave steepness, as expressed by the ratio of wave height to wave length, H/L, and wave age as expressed by the ratio of wave velocity to wind velocity, C/V. Observed values shown by circles. Assumed relationship shown by solid lines. (From U. S. Naval Oceanographic Office.[6])

for H in feet and F in nautical miles. Another early formula taking into account only the wind speed (in knots) is:[6]

$$H_{max} = .026V^2 \qquad (3.7)$$

A rule of thumb is that the highest waves in feet will be approximately 80% of the wind speed in knots, and that swell lose approximately one-third their height each time they travel a distance in miles equal to their length in feet. A more useful empirical relationship for estimating the maximum waves at a site is:[8]

$$H = .0555 \sqrt{V^2 F} \qquad (3.8)$$

and for the maximum period:

$$T = 0.5 \sqrt[4]{V^2 F} \qquad (3.9)$$

with units as previously defined. These relations are particularly use-ful for short fetches and high wind speeds such as found in hurricanes and other intense storms. In shallow water wave generation is again limited by water depth, bottom friction, and percolation in sandy bottoms.

Wave Statistics and Sea Spectra

Because of the random nature of waves, any detailed analysis of the sea state must necessarily involve statistical techniques. Ocean engineers generally analyze wave records by the evaluation of the probability distribution of wave heights to determine the mean, significant height, highest waves, etc., and by use of sea spectra, which gives the entire range of heights versus frequencies (or periods) for a given sea or storm condition. Engineers may additionally be concerned with the probability of encounter or return period analysis for a given wave height. This latter aspect of statistical analysis will be discussed in Section 3.4 in connection with the selection of a design wave.

It has already been mentioned that for a given sea state or wave record, the probability of occurrence of a given wave height (H_i) very nearly follows the Rayleigh distribution. This was first demonstrated by Longuet-Higgins in 1952.[9] The Rayleigh distribution function, shown schematically in Fig. 3.5, is given by:

$$P(H_i) = \frac{2H_i e^{-H_i^2/\bar{H}^2}}{\bar{H}^2} \tag{3.10}$$

where $\bar{H}$ is the root mean square height (Hrms) given by:

$$\bar{H}^2 = \frac{1}{N} \sum_{i=1}^{i=N} H_i^2 \tag{3.11}$$

where N is the number of waves in a record. Naturally the greater the number of waves passing a point the greater are the chances of a wave of extreme height occurring. However, for engineering purposes it is usually considered sufficient that $N = 1000$ and perhaps less, depending upon the purpose.

By integrating the Rayleigh function a cumulative probability distribution can be obtained which yields the given fraction of waves less than or equal to H_i. This has been done in terms of the significant wave height to yield the values given in Table 3.1. It should be borne in mind, however, that the elevation of the sea surface at any point in time at a given location will follow the normal distribution (or bell curve) — hence the reference to a Gaussian sea state. Further, the average wave period, significant wave period, and the period of peak energy are all nearly equal, giving a symmetrical bell curve distribution for the wave period spectrum shown in Fig. 3.4. Note that the significant period (T_s) is taken as the average period of the highest one-third waves. If we consider that wave energy is proportional to the square of the wave height (see Section 3.5) as given by:

$$E = \frac{\gamma}{8} H^2$$

where E is the energy per unit sea surface area and γ is the unit weight of sea water, then the total energy in a sea wave is given by the sum of the many component heights (h_i):

$$E_T = \frac{\gamma}{8} \sum_{i=1}^{i=N} (h_1 + h_2 + \ldots h_N) .$$

If the sum of the squares of the component heights is plotted against the range of periods (or frequency) for the heights within that given frequency range, the result is an energy spectrum for the given sea condition. The seaway can thus be characterized by the energy spectrum, which indicates the amount of energy in the infinite number of component waves that yield the final irregular sea pattern.

Sea spectra are commonly plotted with relative energy density in energy-seconds or feet squared-second units versus circular frequency (ω) where $\omega = 2\pi/T$ in inverse seconds, although sometimes the cyclic frequency (f) is used, where $f = 1/T$. If we define the spectral density, $S(\omega)$, as a function such that any increment of area under its graph, multiplied by a constant, gives the wave energy in that incremental bandwidth of frequency, then the total energy in any increment $\partial\omega$ at the central frequency (ω_n) is given by:

$$\rho g\, S(\omega_n)\, \partial\omega$$

where $\rho = \gamma/g$, and the total energy of the system is:

$$E_T = \rho g \int_o^N S(\omega)\partial\omega \qquad (3.12)$$

Theory shows that the variance, or mean square value, is equal to the area under the spectrum. This implies that if the spectrum is known, then the RMS (root-mean-square) value of the seaway record can be readily determined.

Direct application of sea spectra to problems of ship motions was first introduced by St. Denis and Pierson.[10] The theory of vessel motions (seakeeping) is beyond the scope of this text; interested readers should refer to basic naval architectural texts such as references 11 and 12. Applications of sea spectra to the evaluation of forces and moments on sea structures will be discussed further in Section 3.6. Michel[13] has simplified the explanation of the use of spectra for engineering purposes for those wishing an elementary review of the subject.

Figure 3.11 shows schematically the development of a sea showing the shift from the high frequency toward the low frequency end of the spectrum with increasing wind duration until an FDS is reached and no further increase in energy is imparted to the wave.

Two mathematical forms of sea spectra are commonly used for engineering analysis: the Pierson-Moskowitz spectrum[14] and the Bretschneider spectrum.[15] The use of the PM spectrum is well covered in references 11 and 12 and will not be discussed further here. The Bretschneider spectrum[15] is probably easier to use and

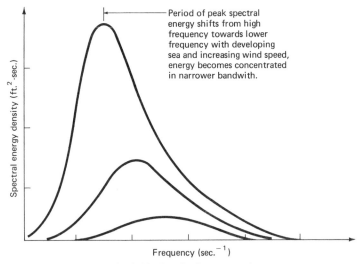

Fig. 3.11. Developing sea spectra.

more practical for structural design applications.   This spectrum is based on the Rayleigh period and height probability distributions assuming zero correlation between individual periods and heights, and is, in the frequency domain, given by:

$$S(\omega) = 2.7 \frac{H_s^2}{\omega} \sqrt[4]{\frac{\omega T_s}{2\pi}}\; e^{-.675\sqrt[4]{\frac{\omega T_s}{2\pi}}} \qquad (3.13)$$

where the significant wave height is found from:

$$H_s^2 = - \int_0^\infty S(\omega)\partial\omega \qquad (3.14)$$

Figure 3.12 represents an example plot of this spectrum for a one-foot significant wave height and a range of significant periods.

Forecasting Methods

The prediction of the spectrum of waves that will occur at a given location and under given conditions is most complex; it still remains as much an art as a science, and the problem is best left to the specialists in this field.   The designer of sea structures, however, must have a grasp of the basics in order to properly interact with such specialists when

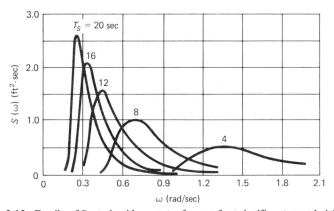

Fig. 3.12. Family of Bretschneider spectra for one-foot significant wave height.

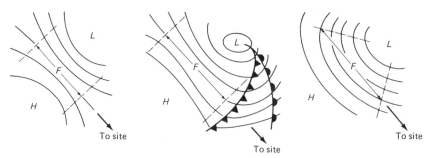

Fig. 3.13. Delineation of fetch for various types of isobars. (After U. S. Naval Oceanographic Office.[1])

attempting to determine the design wave or design wave spectrum. The problem of wave prediction is further complicated by the variability of meteorological conditions, the interaction of swell from remote generating areas with the local storm conditions, the changes that such swell undergo after leaving a generating area, and the change in deep water waves as they move into or are generated over shallow waters (i.e., the effects of shoaling, refraction and diffraction, etc., which are discussed in the next section of this chapter). Wave prediction requires relatively detailed meteorological information such as the intensity, extent, direction, and speed of forward motion, isobar spacing, and so on, so that fetches can be defined and wind velocities determined. Figure 3.13 shows fetches defined for various types of isobars (lines of equal atmospheric pressure on a weather map) and orientations to the coast. For very intense storms such as hurricanes, FDS conditions are not attained because of the required fetches and durations for the high wind speeds.

Bretschneider presented isolines of relative significant wave heights for a slowly moving hurricane,[15] as shown in Fig. 3.14. Once the significant wave height has been determined at some point in the storm as by equation (3.15), then the total wave field can be predicted, as evidenced from the figure. Based on an analysis of 13 East Coast hurricanes, Bretschneider[8] derived the following formula for finding the value of the significant wave height (H_{smax}) at the point of maximum wind corresponding to unity in Fig. 3.14:

$$H_{smax} = 16.5 \, e^{\frac{R\triangle p}{100}} \left[1 + \frac{\alpha 208 V_F}{\sqrt{V_R}}\right] \qquad (3.15)$$

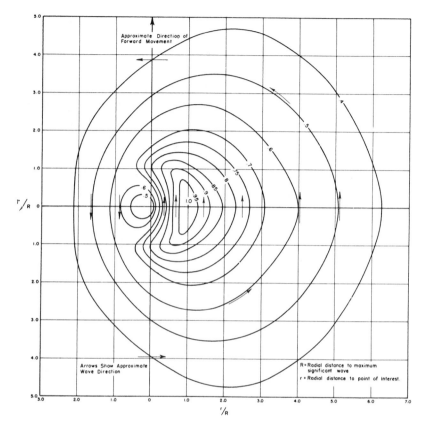

Fig. 3.14. Isolines of relative significant wave height for slowly moving hurricane. (After Bretschneider[15])

where R is the radius for maximum wind (nautical miles) measured from the storm center, ΔP is the reduction of atmospheric pressure at the storm center (inches of mercury), V_F is the hurricane forward speed (knots), α is a factor to account for the effect of the storm forward speed relative to a stationary storm ($\alpha = 1.0$ for slow-moving hurricanes), and V_R is the maximum wind speed (knots) at distance R. A similar relation has been developed for the significant period; however, it can be estimated reasonably well from the simpler formula:

$$T_S = 2.13 \sqrt{H_S}$$

Figure 3.14 was based on the SPH introduced in Chapter 2.2. Either the SPH, PMH, or other storm criteria can be used to define the wind field and generating area for a given locality. Recently Bretschneider and Tamaye[16] have furthered the work on hurricane wind and wave forecasting techniques and present some actual storm spectra as well as nondimensional graphs for determining the wave field. The authors emphasize the importance of considering the total envelope of spectra for a location and storm condition and not just the spectrum at the storm's peak, noting that neglecting the high frequency end of the spectra may grossly underestimate the total energy content.

It is interesting to compare the SPH with the Standard Project Northeaster, a localized extratropical cyclone wind pattern, which is discussed below and in Section 5.2. Although it is a much less intense storm than the SPH, it has a rather large generating area and may remain nearly stationary for up to a few days, thus allowing larger waves to build. Therefore, high wind speeds may not be the most important factor in determining the wave heights and steepnesses in a storm. The storm's extent, speed, and direction of motion, relation to land, and so forth, may be more important. Bretschneider[8] demonstrated the importance of a hurricane's forward speed upon wave growth. Wilson[17] has presented a graphical method of forecasting waves in moving fetches.

There are two basic generalized methods of deep water wave forecasting presently in use: the method originally devised by Sverdrup and Munk in 1947,[18] and modified by Bretschneider in 1952[5] and again in 1958[19] (so that this method is referred to as the SMB method); and the method of Pierson, Neumann, and James, presented in 1955[20] (referred to as the PNJ method). Both methods are popular, but the SMB method is better adapted to engineering purposes.

The PNJ method uses the Neumann energy spectrum[21] and the Rayleigh probability distribution to derive co-cumulative spectra (CCS) curves for various wind speeds, fetches, and durations. Examples of these curves are shown in Figs. 3.15 and 3.16.[20] The CCS is actually the area under the energy frequency spectrum. The use of the curves will become apparent with some perusal. The PNJ and SMB methods were based on different observed data and, therefore, give slightly different results, depending upon the condition.

The forecasting curves in Fig. 3.17 were based upon the SMB meth-od. By considering that the wave height and celerity are both functions of the wind speed (V), duration (t), fetch length (F), and gravity (g):

$$C = f (V, F, t, g)$$
$$H = f (V, F, t, g)$$

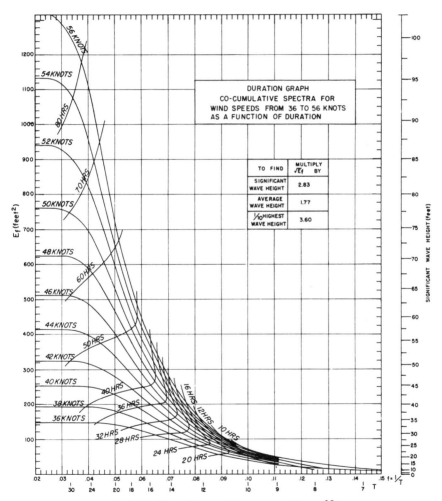

Fig. 3.15. (From Pierson, Neumann, and James[20])

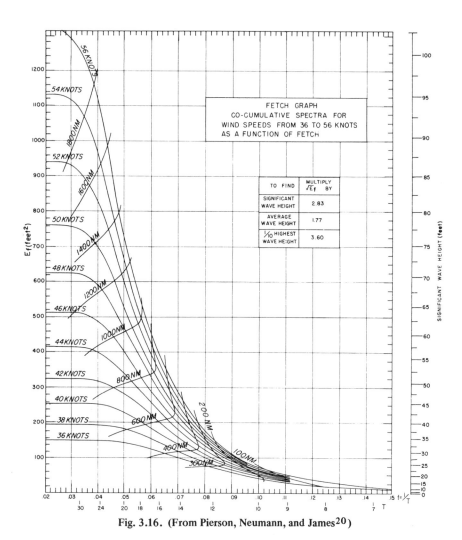

Fig. 3.16. (From Pierson, Neumann, and James[20])

and using the principles of dimensional analysis of the Buckingham-π theorem, as described in most fluid mechanics texts, the nondimensional quantities gH/V^2, C/V, gF/V^2, and gt/V were derived. Wave data from various sources were plotted against these nondimensional parameters, and empirical curves were fitted to the data to yield the forecasting curves shown in Fig. 3.17,[22] which can be read directly to find the deep water significant wave heights and periods as functions of wind speed, fetch, and duration.

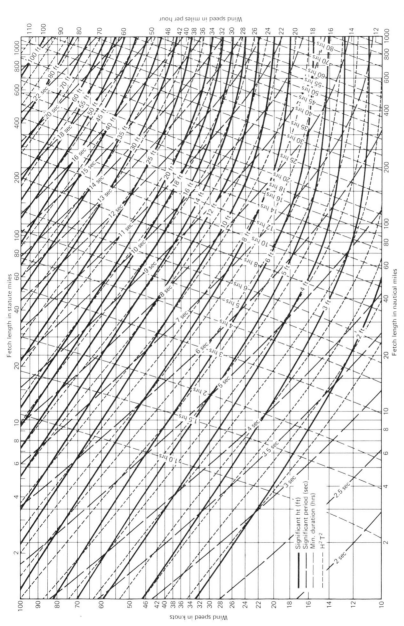

Fig. 3.17. Deep water wave forecasting curves as a function of wind speed, fetch length, and wind duration, for fetches 1 to 1000 miles. (From U.S. Army, CERC SPM.[22])

Kinsman[23] and Bretschneider in Ippen[24] discuss this procedure and the relative merit of both methods in some detail and additionally give preferred methods of calculating wave heights for various types of storm conditions (e.g., moving vs. stationary storm systems).

Shallow water forecasting curves are given in Fig. 3.18,[22] and are presented in nondimensional form, so that the nondimensional parameters gF/V^2 and gd/V^2 where d is the water depth, must first be calculated in order to enter the graph and find the nondimensional wave height gH/V^2. These curves are valid for predicting wind-generated wave heights over bottoms of constant depth as in lakes, bays, and harbors, or along the shallow continental shelves where the waves are not propagating toward shore from deeper water. Waves generated in relatively shallow water will generally be smaller in height and shorter in period than waves generated in deeper water where deep and shallow water are defined relative to the wave length and water depth as discussed in Section 3.5. In general, waves will "feel bottom" when the parameter d/T^2 is less than 2.5 ft/sec^2, and the water depth must be considered as a factor affecting wave growth. Useful graphs for the prediction of wave heights and periods in shallow water of constant depth and over sloping bottoms can be found in reference 22. These curves are based upon the SMB method for deep water wave generation but have also taken into account the effects of bottom friction and percolation as discussed in Section 3.3.

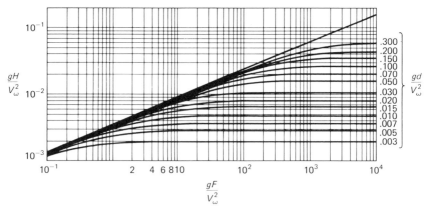

Fig. 3.18. Wave forecasting curves for shallow water of constant depth. (After U. S. Army, CERC, SPM.[22])

Variable bottom topography, restricted fetches, refraction, and interaction of sea and swell from remote storms all affect the spectrum of waves at a given site, so that the application of most of the techniques introduced in this chapter involves the use of computers and/or rigorous computations that are in general beyond the scope of expertise of most structural designers. Although development of sea spectra and prediction of waves is perhaps best left to the specialist, the designer can better interact with the specialist and interpret such information by understanding some of the basics. Some simple relations and tables and graphs have been presented to give the reader a feel for what kind of waves can be developed under given circumstances and for preliminary estimates of possible waves at a site. Interpretation of such preliminary estimates must be applied cautiously, considering the many interacting phenomena and possible conditions. As a matter of practical application, the designer is more concerned with selecting a design wave or design wave spectrum that can be used to evaluate forces and moments; so a return-period-type analysis for extreme waves is of great interest. This aspect of applied wave statistics will be discussed in Section 3.4, while the application of wave spectra is discussed in Sections 3.4 and 3.6.

3.3 MODIFICATION OF WAVE PROPERTIES

General

As waves propagate shoreward or over shallow water, their properties become modified in various ways by interaction with the seabed and ocean boundaries. The wave crests may be bent or distorted by interaction with the bottom contours (refraction), or by passage around or through a fixed barrier (diffraction), and waves may be reflected by steep shorelines or seawalls. All of these processes generally involve a change in wave height, length, celerity, and so on, which must be understood before wave heights at coastal sites can be forecast. Shoaling wave heights are further modified by bottom friction and percolation in sandy bottoms. Wave heights and steepness (the properties generally of most concern to coastal engineers) are further modified by interaction with currents and other waves. Locally the time and height of tide may have a pronounced effect on wave patterns and development. Waves impinging upon gradually sloping shorelines spill over and break, forming surf. Breaking wave run-up and surf zone dynamics pose further problems for coastal engineers. The various interactions just introduced will now be discussed individually, but it must be remembered that many of these effects may be interacting simultaneously. Such complex interactions are normally beyond the scope of the structural designer, as are the traditional coastal engineering problems of littoral transport and sedimentation processes; but these problem areas are introduced herein for completeness.

Effects of Shoaling

The change in wave height for a shoaling wave can be represented by:

$$\frac{H}{H_0} = K_R \; K_s \; K_{FP} \qquad (3.16)$$

where H is the modified wave height and H_0 is the initial deep water height, and K_R, K_s, and K_{FP} are the refraction, shoaling, and friction–percolation coefficients, respectively. The refraction coefficient, K_R, will be discussed in the following subsection. The shoaling coefficient accounts for the effect of change in water depth and is given by:

$$K_s = \sqrt{\frac{n_0 C_0}{nC}} \qquad (3.17)$$

where C and C_0 are the modified and initial wave celerity, and n and n_0 are the modified and initial transmission coefficients, given by linear wave theory (see Section 3.5) as:

$$n = \frac{1}{2} \left[1 + \frac{2 \, kd}{\sinh 2 \, kd} \right] \qquad (3.18)$$

where $k = 2\pi/L$ is known as the wave number. The shallow water wave celerity is related to the deep water celerity by:

$$C = C_0 \tanh kd \qquad (3.19)$$

Wiegel[25] has presented tables for obtaining values of K_s in terms of d/L and d/L_0 where K_s is given in terms of H/H_0. For relative water depths as defined by the parameter d/T^2 greater than approximately 2.5 ft/sec², K_s is essentially equal to unity. For a relatively small range of relative water depths over which the waves first begin to feel bottom, approximately bounded by $d/T^2 < 2.5$ ft/sec^2, but $> .30$ ft/sec^2, K_s drops to a minimum value, and the wave height is thus temporarily reduced. Thenceforth K_s rapidly increases with decreasing depth as indicated in Table 3.3, which gives values of K_s versus the period parameter, d/T^2.

The friction– percolation coefficient is defined as:

$$K_{FP} = \sqrt{\frac{Pb}{P_0 b_0}} \qquad (3.20)$$

where P is the power transmitted across a unit of area of width b and height d, and b is the orthogonal spacing, to be discussed in the following subsection. P_0 and b_0 are the power and orthogonal spacing initially. The K_{FP} is actually the product of the coefficients of bottom friction (K_F) and percolation (K_P), which are normally lumped together. For impermeable bottoms K_P can be taken as unity. For permeable or sandy bottoms where water motion at the bottom interacts with the liquefied sand grains, the K_P may be significant.[26] Based upon theoretical considerations and upon extensive wind and wave data taken by the U. S. Army Corps of Engineers at Lake Okeechobee, Florida, and some ordinary wave data from the Gulf of Mexico,

Table 3.3. Values of the shoaling coefficient, $K_s = H/H_o$, vs. the period parameter, d/T^2.

d/T^2	K_s	d/T^2	K_s	d/T^2	K_s
10	1.000	1.5	0.946	0.3	0.997
5	1.000	1.4	0.941	0.2	1.068
2.5	0.990	1.3	0.934	0.19	1.079
2.4	0.987	1.2	0.928	0.18	1.091
2.3	0.986	1.1	0.917	0.17	1.102
2.2	0.982	1.0	0.917	0.16	1.117
2.1	0.979	0.9	0.914	0.15	1.130
2.0	0.974	0.8	0.913	0.14	1.147
1.9	0.970	0.7	0.915	0.13	1.164
1.8	0.964	0.6	0.922	0.12	1.185
1.7	0.959	0.5	0.935	0.11	1.207
1.6	0.953	0.4	0.957	0.10	1.233

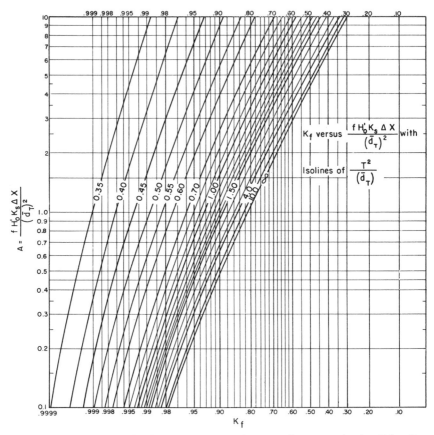

Fig. 3.19.   Relationship for friction loss over a bottom of constant depth.   (After Bret-schneider and Reid[26])

Bretschneider and Reid[26] developed a numerical procedure for computing the generation of wind waves in shallow water of constant depth.   Figure 3.19, based upon their work, gives values of K_{FP} for impermeable bottoms.   A friction factor of $f = .01$ is ordinarily used in computations for shallow water generation, although this factor varies with bottom roughness, and its application is somewhat a matter of judgment.   In Fig. 3.19, ΔX represents the incremental distance over which the reduction in wave height takes place.   This figure can be applied to bottoms of variable depth if an appropriate number of incremental depths are taken, over which the depth is assumed constant.   Those wishing a more rigorous derivation of K_s

and K_P and examples of their application to shallow water wave forecasting techniques, are referred to Bretschneider in Ippen.[24]

Refraction

Just as light rays are bent when traveling from one medium to another, water waves are distorted by changes in water depth in relatively
shallow water. This bending of wave crests, or fronts, is called refraction. If a series of long-crested, regular waves approaches a coastline at some oblique angle, and the bottom contours are relatively
uniform, the portion of the wave crest nearest the shore will feel the
bottom first and thus be retarded relative to the portion of the wave
crest in deeper water so that there will be a bending of the wave
crest along its length. In general, the wave crests tend to make themselves parallel with the bottom contours, as shown in Fig. 3.20. The
fact that wave fronts are not typically long-crested or regular results
in some complex changes in the entire wave spectrum as the lower-

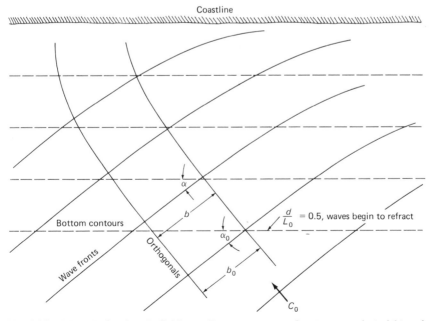

Fig. 3.20.   Wave Refraction Definitions.   Deep water wave fronts assumed straight and
parallel.

frequency component waves will be retarded in deeper water. The following discussion, however, is based upon the somewhat idealized condition of monochromatic, long-crested, regular wave trains. The bending of wave fronts is governed by Snell's law, which states that where the bottom contours are straight and parallel, the sine of the angle (α) between the incident crest and the bottom contour is proportioned to the velocity of wave propagation (C), accordingly:

$$\frac{C}{C_0} = \frac{\sin \alpha}{\sin \alpha_0} \tag{3.21}$$

Referring again to Fig. 3.20 for definitions, an orthogonal is a line everywhere perpendicular to the wave fronts and hence parallel to the direction of wave travel. It is normally assumed that the wave energy contained between orthogonals remains constant as the wave front progresses, which implies that there is no dispersion of energy along the crest, no reflection of energy back from the rising bottom, nor is there loss of energy by any other process. Based on the foregoing discussion and on the fact that wave energy is proportional to the square of the wave height it can be shown that:

$$\frac{H}{H_0} = \sqrt{\frac{b_0}{b}} \; \sqrt{\frac{C_{G0}}{C_G}} \tag{3.22}$$

where b and b_0 are the modified and initial orthogonal spacing and C_G and C_{G0} are the modified and initial group celerity. The refraction coefficient, K_R, then is given by:

$$K_R = \sqrt{\frac{b_0}{b}} \tag{3.23}$$

In accordance with linear wave theory K_R depends only upon wave period, depth contours, and initial wave direction. In general, the orthogonals will tend to either converge or diverge as the wave crests change direction. Converging orthogonals result in an increased concentration of wave energy and hence an increase in wave height such as when waves travel over a shoal, whereas diverging orthogonals decrease the energy per unit area and cause a reduction in wave height, as illustrated in Fig. 3.21. The refraction coefficient is usually obtained from the construction of a refraction diagram. Refraction diagrams are of basic importance to coastal engineering studies, especially in

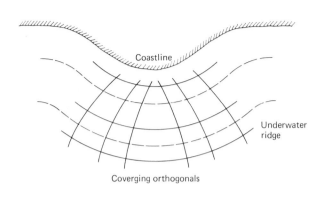

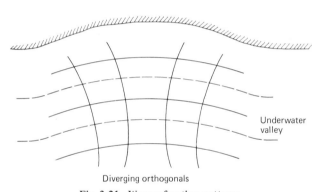

Fig. 3.21. Wave refraction patterns.

the siting of shore protection structures and in the study of littoral processes. These diagrams show the wave fronts, orthogonals, and reduction in height for a given incident wave period, height, and direction. Figure 3.22, from reference 27, can be used to determine the change in wave height and direction for beaches with uniformly sloping bottoms.

There are two general methods used to construct refraction diagrams: the wave front method and the direct ray or orthogonal method. Both methods, which are described in detail in reference 27, yield similar results; however, the direct orthogonal method is more amenable to engineering applications. Nautical charts, of suitable scale, upon which depth contours can be drawn are the starting point for either method. In general, refraction diagrams should start with

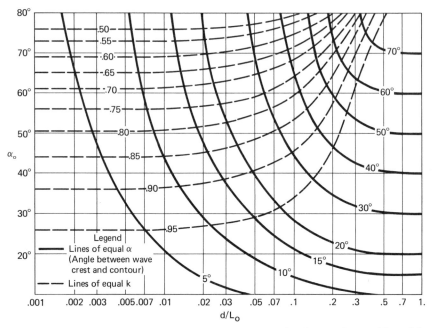

Fig. 3.22.   Change in wave direction and height due to refraction on slopes with straight, parallel depth contours. (After Johnson et. al.[27])

straight wave crests in a water depth equal to half the deep water wavelength, or where $d = 1/2(5.12)T^2 = 2.56T^2$. Normally there will be about as many bottom contours required as the period in seconds of the longest wave to be studied. As the procedure for the construction of refraction diagrams is fairly tedious, the reader is referred to references 22 and 27 or almost any of the basic coastal engineering texts, such as references 28 and 29. Reference 27 includes facsimiles of template devices that can be constructed to facilitate the diagram construction. As may be surmised, a thorough study of any site will require the construction of many diagrams to cover all important wave periods and directions. Wave fronts may also be refracted by currents, as will be discussed later in this chapter.

Diffraction

When waves pass around or through a rigid, impermeable barrier, there is a consequent modification of the wave pattern and change in height,

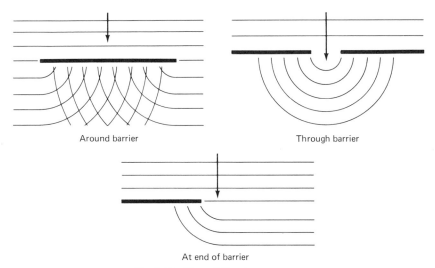

Around barrier   Through barrier

At end of barrier

Fig. 3.23. Wave diffraction patterns.

usually a diminution, as a result of the transfer of energy along the crest into the lee of the structure. Figure 3.23 shows various diffraction patterns schematically. The relative wave height at any point is given by the diffraction coefficient K_D, where $K_D = H/H_0$. Figure 3.24 is a definition sketch for the diffraction of waves behind the end of a semi-infinite barrier. The diffraction coefficient is a function of the angle of incidence, θ, the angle between the barrier and a line connecting the end of the barrier with the desired location, β, and the ratio r/L, where r is the radius vector and L the incident wavelength. Wiegel[30] presented values of K_D in terms of θ, β, and r/L, based upon the work of Penny and Price,[31] who demonstrated that the mathematical solution for the diffraction of light waves can be applied to water-wave crest patterns and height reductions as well. Table 3.4 is extracted from the more complete tabulation presented by Wiegel.

Johnson[32,33] studied diffraction of waves through breakwater gaps and presented plots of wave crest patterns and height reductions for various gap width to incident wavelength ratios, such as the one reproduced in Fig. 3.25.

The problem of waves propagating around a barrier becomes somewhat complicated because of the complex interference pattern that is set up. Likewise, the problems of combined refraction and diffraction effects and of reflected wave patterns are much more complex.

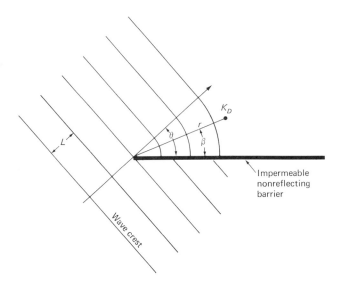

Fig. 3.24. Wave diffraction behind barrier.

Although combined refraction/diffraction effects are a common problem, there are not to date any wholly satisfactory techniques for predicting resulting wave patterns. Model tests are required for any major installation. Reference 22 gives a procedure for dealing with this problem and is a good general reference for solving practical diffraction problems.

Reflection

When a wave encounters a boundary of any kind, some portion of the wave's energy will be reflected back away from the boundary provided the water depth is deep enough that the wave does not break. Wave reflection is more pronounced the more vertical, rigid, and frictionless the boundary is upon which the wave impinges. Taking as an example a vertical, smooth concrete seawall in relatively deep water, nearly all of the wave energy will be reflected seaward again. If the wave front is nearly parallel to the wall, then a standing wave pattern will result in front of the wall with a nodal point at some distance from the wall where the water level remains relatively constant. The standing wave seaward of the nodal point (the antinode) may be up to twice the height of the incident wave. Below the nodal point the water particle motion is nearly horizontal, whereas below the anti-nodes the particle motion will be vertical. The geometry of such a standing wave, called a clapotis, is discussed further in connection with wave forces in Section 3.6. In general, as the wall slope becomes

Table 3.4. Wave diffraction coefficients (K_D).*

Θ	r/L	β 0°	30°	60°	90°	120°	150°	180°
30°	.5	.61	.68	.87	1.03	1.03	.99	1.00
	1	.50	.63	.95	1.05	.98	1.01	
	5	.27	.55	1.04	1.02	.99	1.01	
	10	.20	.54	1.06	.99	1.00	1.00	
60°	.5	.40	.45	.60	.85	1.04	1.03	1.00
	1	.31	.36	.57	.96	1.06	.98	
	5	.14	.18	.53	1.04	1.03	.99	
	10	.10	.13	.52	1.07	.98	1.00	
90°	.5	.31	.33	.41	.59	.85	1.03	1.00
	1	.22	.24	.33	.56	.96	1.05	
	5	.10	.11	.16	.53	1.04	1.02	
	10	.07	.08	.13	.52	1.07	.99	
120°	.5	.25	.27	.31	.41	.60	.87	1.00
	1	.18	.19	.23	.33	.57	.95	
	5	.08	.08	.11	.16	.53	1.04	
	10	.06	.06	.07	.13	.52	1.06	
150°	.5	.23	.24	.27	.33	.45	.68	1.00
	1	.16	.17	.19	.24	.36	.63	
	5	.07	.08	.08	.11	.18	.55	
	10	.05	.05	.06	.08	.13	.54	
180°	.5	.20	.23	.25	.31	.40	.61	1.00
	1	.10	.16	.18	.22	.31	.50	
	5	.02	.07	.07	.10	.14	.27	
	10	.01	.05	.06	.07	.10	.20	

*See Fig. 3.24 for definition of terms. After Wiegel.[30]

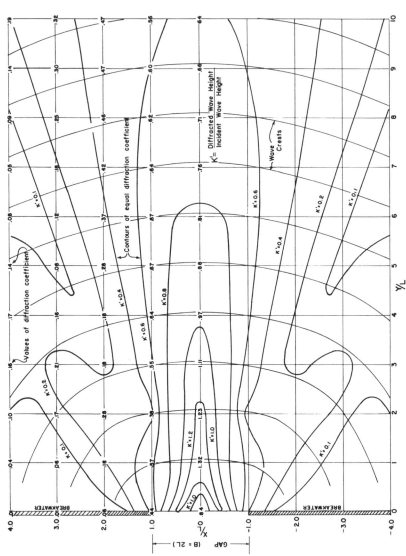

Fig. 3.25. Generalized diffraction diagram for a breakwater gap width of two wavelengths (B/L = 2). (After Johnson[32])

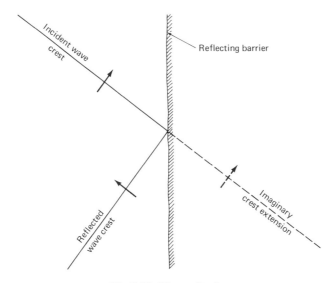

Fig. 3.26. Wave reflection.

more gradual, and the wall roughness and permeability increase, the height of the reflected wave will decrease. Steeper waves are reflected less than flatter waves.

If the wave, or wave train, strikes the boundary at some oblique angle such as shown in Fig. 3.26, then the wave front will be reflected at approximately the same angle (i.e., the angle of incidence equals the angle of reflection). A reflection coefficient (C_R) can be assigned to a given obstacle or barrier where $C_R = H_R/H_0$. Wave reflection within a harbor can be a very menacing problem. In general it is desirable to destroy wave energy within a harbor by reducing reflection at the boundaries. This problem is discussed further in Section 3.7.

Effects of Floating and Submerged Barriers

For a deep water wave ($d/L \geqslant .5$), over 70% of its kinetic energy is concentrated in the upper 20% of the water column, as indicated in Fig. 3.27, which was redrawn from the reference 34. This presents a basic rationale for the use of floating breakwaters. Referring to Fig. 3.28 for a rigidly moored barrier, both a reflection (C_R) and a transmission coefficient (C_T) can be defined, where $C_R = H_R/H_0$ and

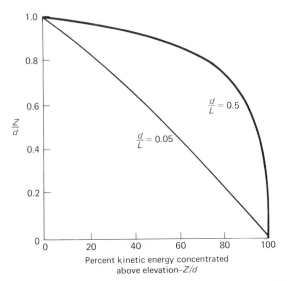

Fig. 3.27. Vertical distribution of kinetic energy in a wave. (After Dean and Harleman.[34])

$C_T = H_T/H_0$. If energy dissipation is neglected, we can relate these coefficients by:

$$C_T = \sqrt{1 - C_R^2} \qquad (3.24)$$

Therefore, C_T can be closely estimated when C_R has been determined. In general, as the incident wavelength increases, C_R decreases while C_T increases. The extent of the structure perpendicular to the wave

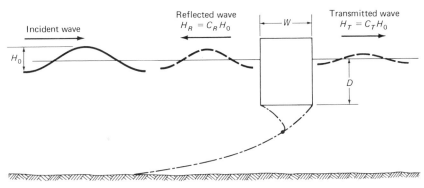

Fig. 3.28. Wave reflection and transmission at moored barrier. (Adapted after Dean and Harleman[34])

direction, relative to the wavelength, is very important, since when the wavelength exceeds about ten times the structure's length, the structure becomes ineffective in reflecting wave energy (in other words, it becomes "wave-transparent"). In relatively shallow water ($d/L < .5$), the wave's kinetic energy is more uniformly distributed throughout the height of the water column, and floating breakwaters become less effective at reflecting wave energy. The C_R for a floating breakwater also depends upon how rigidly the breakwater is moored, as the breakwater's wave-induced motion tends to aid in the transmission of wave energy. Kincaid,[35] as reported by Dean and Harleman,[34] studied the effects of natural period on the effectiveness of floating breakwaters and found that C_R near the breakwater's natural period exceed C_R for the rigidly fixed case for the range of periods studied. Floating breakwaters are discussed further in Sections 3.6 and 3.7 in connection with wave forces and wave-induced motions.

Large submerged objects can be treated as a special case of a shoaling wave when situated in shallow water. The evaluation of forces and scouring action on submerged structures are discussed in Sections 3.6 and 3.7.

Effects of Currents

When waves encounter a current moving in the opposite direction, the wave height is increased, the length and celerity are decreased, while the period remains nearly constant. The opposite effects occur when the current is moving in the same direction as the waves. If we consider a wave train with celerity C_0, approaching a harbor entrance or river mouth with a uniform current velocity of U_c, and further consider that the period (T) remains constant before and after encountering the current (it must, as otherwise there would be a steady build-up or disappearance of waves between the sea and river mouth), then we can write:

$$T = \frac{L_0}{C_0} = \frac{L_c}{U_c + C_c}$$

If deep water relations apply, then:

$$\frac{L_c}{L_0} = \frac{C_c^2}{C_0^2}$$

Combining these equations and rearranging to solve for the modified wave celerity (C_c), we obtain:

$$C_c = \frac{C_0}{2}\left[1 + \sqrt{1 + \frac{4U_c}{C_0}}\right] \qquad (3.25)$$

By considering the conservation of wave power, the change in wave height can also be calculated. When the current opposes the wave direction, the wave steepness will build until the wave breaks with increasing U_c. If $U_c \geqslant .25\, C_0$, the wave cannot propagate against the current. Dalrymple and Dean[36] investigated the maximum heights of waves that could be obtained on uniform currents. They presented nondimensional curves that give the maximum height, crest elevation, and breaking limit for given values of L_0, C_0, and U_c. The authors note that the effect of the current is to increase the limiting wave heights when the current flows in the direction of the waves and to increase the percentage of wave crest above still water. The opposite is true for opposing currents. Determination of the maximum wave height that can be borne on a current and of other properties of such a wave may be of critical importance in the selection of a design wave at certain sites. Herbich and Hales[37] have treated the problem of waves in tidal inlets.

If a current intersects a wave train at some oblique angle (i.e., the orthogonals are not parallel to the current direction), the waves will be refracted similarly to the refraction of shoaling waves. Snell's law (equation 3.21) can be applied where α_0 is the angle between the direction of the current and the incident wave crest, as shown in Fig. 3.29. Johnson[38] used this relation to solve for the change in wavelength due to refraction and found:

$$\frac{L_0}{L_c} = \left[1 - \frac{U}{C_0}\sin \alpha_0\right]^2 \qquad (3.26)$$

The direction and celerity of the modified wave train can then be found from equation (3.21) and the change in height from equation (3.22).

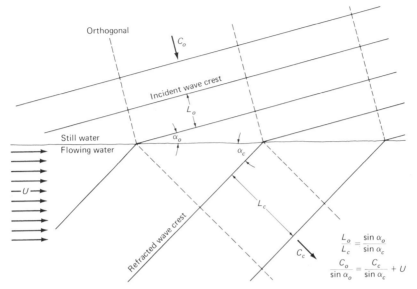

Fig. 3.29. Refraction of waves by current.

Breakers and Surf

As waves propagate into progressively shallower water, the water particle orbits become more elliptical than circular and begin to tilt forward at the crest until the particle velocity at the wave crest equals the wave celerity and the wave begins to break, spilling over and expending its energy in turbulence, bottom friction, and perhaps the formation of a new but smaller wave. Table 3.5 gives coefficients of wave celerity and orbital velocity in shallow water in terms of relative water depth (d/L). Note that the phase velocity is decreasing while the crest particle velocity is increasing. Waves in deep water may also break when they become too steep. The theoretical limit for a wave breaking in any depth of water is given by:

$$\left(\frac{H}{L}\right)_{\max} = \frac{1}{7} \tanh kd \qquad (3.27)$$

In deep water $(d \gg L)$ this reduces to:

$$\left(\frac{H}{L}\right)_{\max} = \frac{1}{7} \qquad (3.28)$$

and in shallow water $(d \ll L)$ this becomes:

Table 3.5. Coefficients for velocity of propagation
(C) and orbital velocity (u), shallow water.

d/L	C	u
.05	.552	1.814
.10	.746	1.340
.15	.858	1.165
.20	.922	1.085
.25	.958	1.044
.30	.977	1.023
.35	.988	1.013
.40	.994	1.007
.45	.997	1.004
.50	.998	1.002
.50	Deep water equations apply	

$$\left(\frac{H}{L}\right)_{max} = \frac{1}{7} kd = \frac{1}{7} \frac{2\pi}{L} d \qquad (3.29)$$

Dividing through and rearranging:

$$\left(\frac{H}{d}\right)_{max} = 0.9$$

Waves normally break, however, when $H/d = .78$ or when they enter a water depth equal to about $1.28H$, where H is the deep water wave height. The theoretical development of the above relations is discussed further in Section 3.5.

The breaking of waves in deep water has a pronounced effect on the development of the wave spectrum, as the shorter-period wave heights and hence the energy density at the higher frequencies are limited by this criterion.

Long period waves, or swell, that pass over an underwater shoal not shallow enough to cause breaking may steepen into a heavy rolling swell known as ground swell.

The slope and roughness of the bottom affect the actual depth and manner in which a wave will break. Referring to Fig. 3.30, if the bottom slope is gradual, a spilling breaker will result. Spilling

Spilling:

Plunging:

Surging:

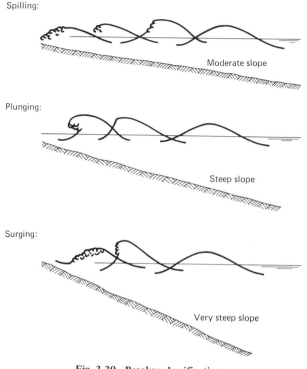

Moderate slope

Steep slope

Very steep slope

Fig. 3.30. Breaker classification.

breakers expend their energy in breaking over long distances. On a steeper slope the wave may curl over suddenly and break in a single plunging crash, called a plunging breaker. Surging breakers occur on very steep slopes and do not appear to actually break, but suddenly peak and dissipate their energy by surging up the beach face. The horizontal extent to which the water from a breaking wave travels is called the limit of up-rush, and the vertical height to which it rises is called the run-up, as illustrated in Fig. 3.7.

The type of breaker to which a structure may be subjected is important in figuring the stability and forces acting upon it. The type of breaker also affects the run-up on the structure. Further, a given structure may be subjected to different types of breaking wave conditions as the tide level changes or the bottom topography is modified by the continual wave action. Figure 3.31, reproduced from reference 22, can be used to determine the breaker height, from the deep water height

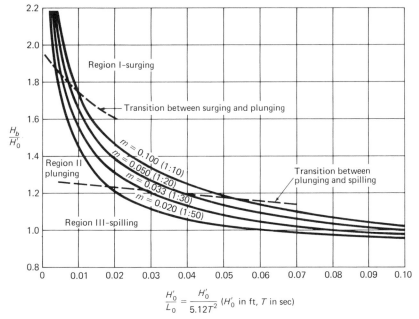

$$\frac{H'_0}{L_0} = \frac{H'_0}{5.127^2} \quad (H'_0 \text{ in ft, } T \text{ in sec})$$

Fig. 3.31.   Breaker height index vs. deep water wave steepness. (From U. S. Army, CERC, SPM.[22])

and length for spilling, plunging, and surging breakers, and Fig. 3.32, from the same reference, can be used to determine the actual depth at breaking for various breaker heights and periods on different slopes. Wiegel and Beebe[39] found the crest elevation (η_c) above the still water level to be about .78H_b, where H_b is the breaker height.

A series of breaking waves is termed surf. Surf can result from swell that has traveled long distances as well as from locally generated wind waves. The energy density in the surf zone is generally much greater than in the storm that originally generated the waves, because of the rapidity with which waves expend their energy during breaking. Waves breaking over an offshore bar may re-form when reentering deeper water, and in some locations an inner and outer line of breakers may be seen. Currents are set up in the surf zone as a return flow of water from the breaking waves. Longshore and rip currents set up by breaking waves are discussed briefly below in Section 4.1. Breaking waves cause a temporary rise in water level along the coast called wave set-up, a phenomenon discussed in Section 5.2. It is often

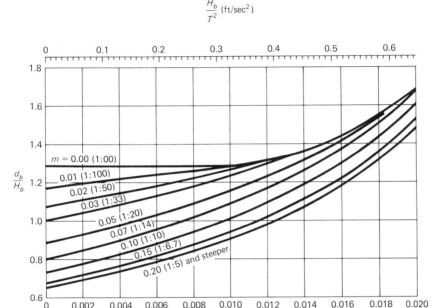

Fig. 3.32.   Dimensionless depth of breaking vs. breaker steepness. (From U. S. Army, CERC, SPM.[22])

observed at beaches that the water level may fluctuate, temporarily increasing when a series of particularly large waves have broken and lowering during intervals of smaller breakers. Sometimes groups of larger and then smaller breakers occur with a definite rhythmic period (usually around 2 to 3 minutes) called surf beat, which is usually due to the interference pattern set up by two or more regular wave trains approaching the beach at the same time.

As previously mentioned, wave action is constantly altering the bottom topography in the surf zone, and the altered bottom contours in turn affect the waves. Surf zone dynamics, littoral processes, and sediment transport are not within the scope of this text. Needless to say, it is generally not feasible or otherwise desirable to construct structures in the surf zone with the exception of special shore protection structures such as groins and jetties. Design of these structures is well covered in reference 22. For a more detailed description of surf zone dynamics and phenomena the reader is referred to reference 40. The problem of wave run-up on structures is treated in Section 3.7.

3.4 THE DESIGN WAVE

Ultimately the evaluation of loads and stresses on sea structures requires the selection of wave parameters, a height and period or range of periods, and/or a complete wave height and period spectrum and the application of an appropriate wave theory to adequately predict water particle velocities and accelerations for the given ambient conditions. The ranges of applicability of the various water wave theories are discussed in the following section. In general, the maximum wave that can occur at a site will be limited by the water depth at the site, the sheltering of the site by the proximity of land, and/or the underwater topography surrounding the site; or the maximum wave for deep water sites will be limited by the low probability of occurrence of the extreme meteorological conditions required for the development of large waves. These limiting conditions will be elaborated upon further in this section.

The application of a design wave height and period to the direct calculation of forces and moments is a process of static analysis. It must be borne in mind, however, that in the case of offshore deepwater structures and at certain shallow water sites, a dynamic analysis considering the structure's response to wave excitation is essential and may govern the structural design. In such cases it is not sufficient to select a design wave, but a complete spectrum or series of storm spectra will be required to determine the frequency content and the structure's response. Further, structures may not fail because of the highest incident wave, but failure may be due to fatigue[41] or to resonant motions[42] set up by some wave trains of considerably less height than the maximum expected wave. Deep water fixed platforms with relatively long periods, on the order of several seconds, are most susceptible in this regard, especially when the structure's fundamental period, torsional period, and/or any of its harmonic modes are nearly attuned to the period of significant or maximum spectral density.

Further, it must be noted that perfect resonance is not required in order for dynamic amplification of forces to take place. Wirsching and Prasthofer[43] have presented a relatively simple algorithm for the preliminary evaluations of deep water fixed platform response to wave loading, which may be helpful in determining the degree of importance of dynamic effects prior to selection of a design wave. Aspects of dynamic response to waves are discussed further in Section 3.6 but have been introduced in this section to emphasize that a single "maximum" design wave may not be sufficient for the design of some structures. In any event, the concept of a design wave remains central to the design of both coastal and offshore structures although there is some difference in philosophy of design wave application.

At present the usual standard practice is first to obtain a long-term probability distribution for the site under consideration. This is usually obtained from a specialist oceanographer or meteorologist familiar with the local climatic conditions and is likely to be presented in terms of the significant wave height. The probabilities of the other heights then can be found through the application of the Rayleigh distribution introduced in Section 3.2. The actual highest wave in relation to the significant wave depends upon the total number of waves in the record, the value approaching two times the significant height for $N =$ 1000 waves. In general, if a significant height is maintained for 6 hours or more, the maximum height will approach twice the significant height. The long-term distribution can be obtained from hindcasting of historical storm records and/or be predicted on probability of occurrence of meteorological conditions required for such wave development. The long-term distribution is fitted to whatever data are available and extrapolated to the 50- or 100-year return period probability level. The accuracy of this extrapolation is assessed by how well the data fit a straight line when plotted on probability paper. Figure 3.33, reproduced from reference 44, is an example of such long-term distributions for several areas of current interest in offshore oil drilling.

These distributions are presented in the form of a Weibull distribution function,[45] although other distribution functions, such as summarized by Bretschneider,[46] have been used in wave data applications. For a given significant height, spectra can then be generated using the SMB or PM methods introduced in Section 3.2. When the design height and spectrum have been determined, a corresponding period or range of periods may be selected in evaluating forces. The frequency distribution of average wave periods should be obtained for this purpose. The choice of period may be critical, depending upon the water depth at the site and whether drag or inertial force compo-

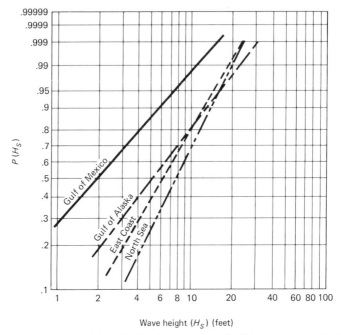

Fig. 3.33. Long-term probability distributions for several offshore areas. (After Riggs.[44])

nents are significant. Reference 47 suggests investigating a range of periods: $\sqrt{6.5H} < T < 20$ for T in seconds and H in meters.

In the design of most coastal and waterfront structures it is usually sufficient to design for a maximum design wave condition with perhaps a range of periods. However, the development of sea spectra and application of spectral analysis techniques is required in the design of floating structures and in the dynamic and fatigue analysis of offshore structures. There are numerous papers on the application of wave statistics to design, of which references 10, 13, and 48 are particularly recommended for the application of wave spectra in the design of floating structures.

In harbors, bays, lakes, and other bodies of water where wave generation is restricted by relatively short fetches and/or shallow water depths, the design wave may be reasonably estimated from the curves in Fig. 3.18. Waves generated over such bodies of water tend to be relatively steep, and the reader should refer to the source of these curves[22] for methods of obtaining corresponding wave periods. Caution must also be exercised to be sure that larger ocean waves may not propagate through the harbor entrance or bay mouth. The design wave or wave spectra for a site may alternatively be predicated upon a hypotheotical design storm such as the SPH or PMH introduced in Section 2.2. Again, such a storm is normally associated with some probability of occurrence. Such storm waves must be propagated into the site area, considering the effects of refraction, shoaling, bottom friction, etc. Table 3.6, reproduced from reference 8, summarizes the highest deep water significant wave heights for selected hurricanes occurring off the U. S. East Coast. This tabulation gives the forecast wave heights for the actual storm motion observed and, for comparison, the heights that would have occurred for a stationary storm of the same intensity and for a storm with critical forward speed resulting in maximum wave generation. The forecasting curves given in Fig. 3.17 are based upon constant wind velocities and straight fetches, which do not obtain within severe tropical revolving storms such as hurricanes.

Actual storm wave data are quite sparse; so references 49 and 50, which report on wave statistics from actual storms, are of particular interest. Sellars,[51] from whose work Table 3.7 is reproduced, has summarized extreme wave conditions for design. Sellars's work covers most world oceans and many years of observation from ocean weather stations, and includes a list of other important references. The maximum possible waves that could occur in the oceans have not yet been recorded; but the meteorological conditions required to produce such waves − of up to and over 200 feet in height! − have been investigated and are possible.[52]

General presentations of ocean wave statistics of particular interest include the work of Hogben and Lumb,[53] Roll,[54] and Yamanouchi and Ogawa.[55] Numerous site-specific studies have been conducted for various areas of the world, in particular those areas of offshore oil drilling interest. Many of these studies are proprietary and generally unavailable. Wave statistical studies for U. S. coastal waters that are readily available include: reference 56, for extreme wave heights along the U. S. Atlantic Coast; reference 57, hurricane wave heights for the Gulf of Mexico; reference 58, Great Lakes wave heights; and reference 59, which summarized coastal wave gage data along the U. S. Atlantic, Pacific, and Gulf coasts over a 20-year period. Refer-

Table 3.6. Summary of highest deep water significant waves predicted for 13 selected hurricanes off eastern coast of United States.

No.	Date	DATA FOR STATIONARY STORM		DATA FOR STORM MOVING AT ACTUAL SPEED		DATA FOR STORM MOVING AT CRITICAL SPEED			RATIO	
		H_O in feet	T_S in seconds	H_{ac} in feet	T_{ac} in seconds	V_{cr} in knots	H_{cr} in feet	T_{cr} in seconds	H_{cr}/H_O	V_{cr}/T_S
(1)	(2)	(3)	(4)	(5)	(6)	(7)	(8)	(9)	(10)	(11)
1	10/15/54	37.9	13.0	59	16	24.8	59.8	16.4	1.58	1.91
2	9/19/55	34.2	12.4	42	13.5	24.1	56.4	15.9	1.65	1.94
3	8/12/55	31.0	11.8	37	13.0	22.7	50.0	15.1	1.61	1.92
4	12/2/25	28.7	11.3	40.5	13.5	22.0	47.7	14.6	1.66	1.95
5	9/3/13	26.0	10.8	37.5	13.0	20.7	42.1	13.7	1.62	1.92
6	8/17/55	23.9	10.3	35.0	12.5	20.0	39.0	13.2	1.63	1.94
7	9/17/06	23.5	10.2	35.5	12.5	19.3	37.2	12.8	1.58	1.89
8	8/11/40	22.6	10.1	31.5	12.0	18.7	34.7	12.4	1.54	1.85
9	8/28/11	22.4	9.9	28.0	11.0	18.6	34.7	12.3	1.55	1.88
10	8/14/53	20.8	9.7	29.0	11.5	18.5	29.5	12.2	1.42	1.91
11	10/15/47	19.5	9.4	28.0	11.0	17.2	28.2	11.4	1.45	1.83
12	8/1/44	19.1	9.3	28.0	11.0	17.5	29.5	11.5	1.55	1.88
13	8/30/52	18.5	9.2	25.0	10.5	17.1	27.6	11.3	1.49	1.86
							Average		1.56	1.90

After Bretschneider.[8]

Table 3.7. Maximum wave heights recorded.

Location	Height ft	Period sec	Steepness	Comment	References
Gulf of Mexico	71.5 (21.8 m)	12.0	0.097	Hurricane Camille wind speed = 120 mph (109 knots) waves over top of wave staff	[30]
North Atlantic Station Juliett	67.0 (20.4 m)	–	–	Occurred 12 Sept., 1961 Tucker gage	[34]
Antarctic	81.7 (24.9 m)	11.0	0.131	Data from sterophoto	[35]
Pacific	80.0 (24.4 m)	16.2	0.060	Data from motion picture film, occurred December 1, 1969	[31]
North Sea	61.0 (18.6 m)	13.3	0.067	Occurred 23 November 1969 Tucker gage	[36]
North Atlantic Station India	65.0 (19.8 m)	12.5	0.081	Occurred February 16 1962 in 44-knot wind Tucker gage	[28]

After Sellars.[51]

ence 60 lists some additional local wave studies and gives guideline wave heights for design of offshore structures in U. S. coastal waters, for depths over 300 feet, from which Table 3.8 is reproduced. The reference level heights given in the table are intended to be applied to a specific set of conditions and, with a specified combination of other provisions of reference 60, to result in a given level of design force and overturning moment on offshore platforms. Information from this table, therefore, should not be applied indiscriminately but with regard to the conditions and provisions of the referenced document.

After the basic deep water wave height has been ascertained, it must be further determined if the water depth at the site is great enough so as not to cause breaking of the highest waves. At many shallow water sites, the largest storm waves may break well seaward of the structure. When considering the effects of wave breaking, one must determine the maximum (and minimum) storm tide (see Section 5.2) levels that may prevail at the time. Figure 3.34 shows two waves with the same deep water properties (H_0, L_0, and T); one of these

Table 3.8. Wave parameters for ten areas in United States waters.

	Wave Height				Wave steepness	Reference level deck clearance*	
	Reference level		Guideline				
	ft	m	ft	m		ft	m
Offshore Gulf of Mexico	70	21.3	(see Fig. 2.8-2)		1/12	48	14.6
Offshore Alaska							
1. Lower Cook Inlet	60	18.3	50–70	15–21	1/13	56	17.1
2. Icy-Bay Gulf of Alaska	100	30.5	90–120	27–37	1/15	80	24.4
3. Kodiak Shelf-Gulf of Alaska	90	27.4	80–110	24–34	1/15	72	21.9
4. Bering Sea/Bristol Bay	85	25.9	75–95	23–29	1/13	63	19.2
Offshore California							
1. Santa Barbara Channel	45	13.7	40–50	12–15	1/16	38	11.6
2. Outer Banks	60	18.3	55–70	17–21	1/15	44	13.4
Offshore Atlantic Coast							
1. Georges Bank	85	25.9	75–95	23–29	1/12	59	18.0
2. Baltimore Canyon	90	27.4	80–100	24–30	1/12	62	18.9
3. Georgia Embayment	75	22.9	65–85	20–26	1/12	53	16.2

*Above M.L.W. in Atlantic; above M.L.L.W. in Pacific.

From API, *Rec. Practice for Planning, Designing, and Constructing Fixed Offshore Platforms.* [60]

waves is shown breaking and the other nonbreaking owing to the increase in storm water level, which may greatly affect the maximum height of wave at a site. Equally important, the lowest water level at a site may cause the same wave to break at or on the structure, possibly resulting in even greater loads.

Waves are normally assumed to break in water of depth $d_b = 1.28H_b$, based upon the solitary wave theory (discussed in Section 3.5) and confirmed somewhat by the work of Wiegel and Beebe.[39] However, the actual depth at breaking may vary in proportion to breaker height and with the bottom slope and roughness and original deep water characteristics. The curves in Figs. 3.31 and 3.32 should, therefore, be used in determining breaker height and depth from the deep water characteristics. Galvin[61] investigated the travel distance of breaking

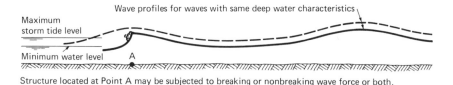

Structure located at Point A may be subjected to breaking or nonbreaking wave force or both, depending upon stage of tide, storm surge elevation, relative wave height, etc.

Fig. 3.34. Example of breaking and nonbreaking waves at same location.

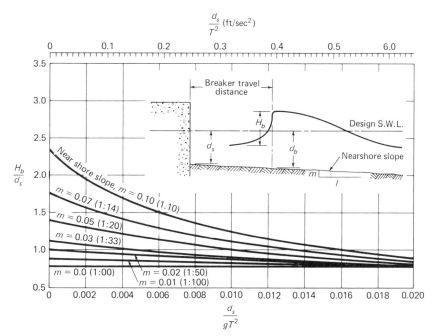

Fig. 3.35. **Dimensionless design breaker height vs. relative depth at structure. (From U. S. Army, CERC, SPM.[22])**

waves and found that for plunging type breakers the horizontal distance over which the wave traveled (the plunge distance) from the beginning of breaking to a point where it became a turbulent bore varied from about two times the breaker height (H_b) to $4H_b$ for beach slopes (m) of about .20 to $m = .05$. For seawalls fronted by beaches of various slopes, breaker height and travel characteristics can be obtained from Fig. 3.35, reproduced from reference 22 and based upon the work of Weggel.[62]

Breaking waves travel with up to 80% or more of their height above the nominal still water level. Offshore the crest elevation may lie in the order of 60% to 70% above the S.W.L., a factor of critical importance when determining the bottom elevation of a fixed platform, for example. The curves in Fig. 3.36 may be used to determine the crest elevation above S.W.L. for waves in any depth of water. Bretschneider[63] has presented a concise summary of the important aspects of crest elevation, height of wave crests at breaking, and other

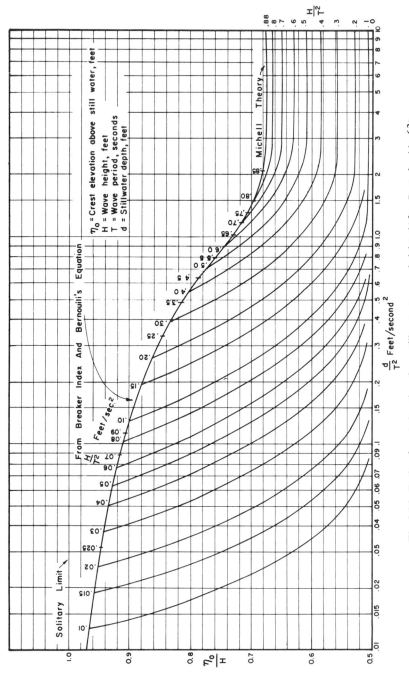

Fig. 3.36.   Ratio of crest elevation above still water to wave height. (After Bretschneider[63])

important criteria in the selection of design waves for offshore structures.

In the design of coastal protection structures, consideration is given to structure type, or function, in determining the design wave height.[22]   The structures may be loosely classified as: rigid, semi-rigid, or flexible. Rigid structures such as bulkheads and seawalls may be subject to catastrophic failure from a single high wave; thus such structures are normally designed for the highest (H_{max}) or highest 1% $(H_{1/100})$ of the waves. Semi-rigid structures, such as sheet pile cells, which yield somewhat and thus absorb some of the wave energy, may be designed for $H_{1/100}$ or $H_{1/10}$; whereas flexible structures such as rubble mound breakwaters are often designed for H_s on the basis that infrequent impingement of higher waves should not cause serious damage. Flexible structures are usually destroyed progressively over long periods of time by normal wave action and the accumulated damage of many storms; so a more complete wave climatology[59] may be of greater value than a single design wave analysis.

In the design of any structure subjected to nondeterministic and probabilistic environmental loads, the ultimate design condition must invariably be weighed against capital cost, maintenance and repair cost, and the consequence of damage with regard to the type of failure. Obviously the possibility of a sudden catastrophic failure, and consequent loss of life and investment, as a result of a single high wave is of much more serious concern than the replacement of a section of armor stone on a breakwater brought about by a single wave with the same likelihood of occurrence. The design of breakwaters and seawalls must also consider some acceptable level of overtopping. This is a functional criterion rather than a structural one, but it nonetheless requires knowledge of the extreme wave climate.

3.5 WAVE THEORIES

General Discussion

Theoretical treatment of water waves, and of sea motion in general, involves relatively rigorous mathematical analysis. The intent of this section is to promote a general understanding of the development of such theory as it relates to describing the physical phenomenon without going into the details of the mathematical manipulations. For a rigorous treatment of free surface hydrodynamics, the reader is referred to the classic work of Lamb.[64] For a more in-depth but general treatment of water wave theories with engineering applications, the reader is referred to the textbooks in references 28, 65, and 66.

All sea motions can be completely described by differences in velocity and pressure as functions of space and time. The basic hydrodynamic equations governing all sea motions are the equations of continuity and momentum. The form and solution of these equations can be varied, depending upon the application and solution sought, but, in general, in all solutions an incompressible, inviscid, and irrotational fluid is assumed. A velocity potential (ϕ), similar in concept to a scalar force potential, such as the gravity potential of classical physics, can be defined and will exist for the assumption of irrotational flow (i.e., there is no net rotation of an individual fluid element, only distortion). The velocity potential is defined as a function whose negative derivatives yield the water particle velocities in each of the coordinate directions (refer to Fig. 3.37 for definitions), such that:

$$u = \frac{\partial \phi}{\partial x}, \ v = \frac{\partial \phi}{\partial y}, \ w = \frac{\partial \phi}{\partial z}$$

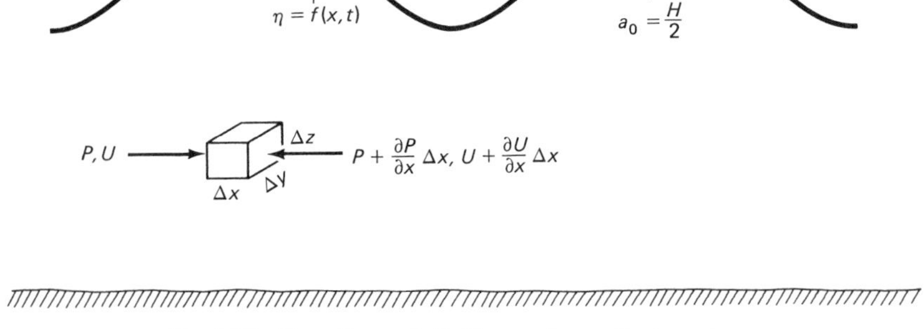

Fig. 3.37. Wave theory definitions and coordinate system.

The equations of continuity and momentum are, in terms of ϕ, in their most general form:

Continuity (the LaPlace equation) –

$$\frac{\partial^2 \phi}{\partial x} + \frac{\partial^2 \phi}{\partial y} + \frac{\partial^2 \phi}{\partial z} = 0 \qquad (3.30)$$

Momentum (the Bernoulli equation) –

$$\frac{\partial \phi}{\partial t} + \frac{1}{2}(u^2 + v^2 + w^2) + \frac{P}{\rho} + gz = f(t) \qquad (3.31)$$

where u, v, and w are the velocity components in the x, y, and z directions, respectively. P is the pressure, and $f(t)$ is an arbitrary function of time which can be incorporated into the term $\partial \phi / \partial t$. The solution of equations (3.30) and (3.31) is carried out as a series expansion in ascending powers of various wave properties for the given boundary conditions. For deep water the expansion is taken in ascending powers of H/L, and in shallow water in ascending powers of H/d. These expansions form the basis of the sinusoidal and cnoidal theories, respectively, to be discussed. The wave is described by the three independent parameters H, L, and d. There are, in general, two boundary conditions – the sea bed, where:

$$z = -d \quad \text{and} \quad w = \frac{\partial \phi}{\partial z} = 0$$

and the free surface (which can be written in terms of either the kinematic or dynamic free surface conditions), where:

$$\frac{P}{\rho} = 0 \quad \text{and} \quad \eta = \frac{1}{g}\frac{\partial \phi}{\partial t} t$$

Other boundary conditions such as a vertical wall or pile, for example, alter the solution. The phase position is given by:

$$\theta = kx - \omega t \qquad (3.32)$$

where k is known as the wave number and is equal to $2\pi/L$ and ω is the angular frequency previously defined as $2\pi/T$. Note that for regular waves:

$$\frac{\omega}{k} = \frac{L}{T} = C$$

Other assumptions usually involved in the evaluation of the hydrodynamic equations are that the bottom is stationary, impermeable, and horizontal; surface tension and capillary effects are negligible; and the pressure along the air–sea interface is constant. The simplest solution involves the further assumption that the waves are of small amplitude; that is, the wave amplitude ($a = H/2$) is small compared to the wavelength (L) and water depth (d). The basic solution for the two-dimensional case of a free, periodic surface wave of small amplitude was first carried out by Airy,[1] and is usually referred to as the first order approximation, the small amplitude wave theory, or the Airy wave theory. An understanding of small amplitude wave theory is basic to coastal and ocean engineering, and to the understanding of higher order theories. Wave theories predict the water particle velocities and accelerations, and the surface profile and pressure at any location, and in general completely describe the water motion. Small amplitude theory, however, does not accurately predict the properties of waves near breaking; so here the solitary wave theory, which assumes a single translating wave crest, is applied. There are many other forms of solution of the basic hydrodynamic equations with respect to the assumed boundary conditions, all of which have preferred ranges of applicability depending upon the problem at hand. Figure 3.38, based on the work of LeMéhaute,[67] shows the relative ranges of application of some of the more prominent wave theories to be discussed.

Note that in Fig. 3.38 deep water has been defined by $d/L > .5$ and shallow water by $d/L < .04$, leaving a transitional zone between these limits. It is in the transitional zone that the engineer must apply some judgment as to which theory best suits the given application. The relative merits of the various theories will be described individually. In the shallow water regime, $d/L < .04$, the choice is mostly limited to the cnoidal or solitary wave, which is in fact a limiting case of cnoidal theory. In deep water, linear (Airy) theory is probably sufficient for most engineering purposes; however, the higher order, Stokes, theories give a better representation of steeper waves. In general, the higher orders tend to predict higher particle velocities, whereas the particle accelerations are not significantly changed. Therefore, in structures where drag forces are predominant, a higher order theory should be

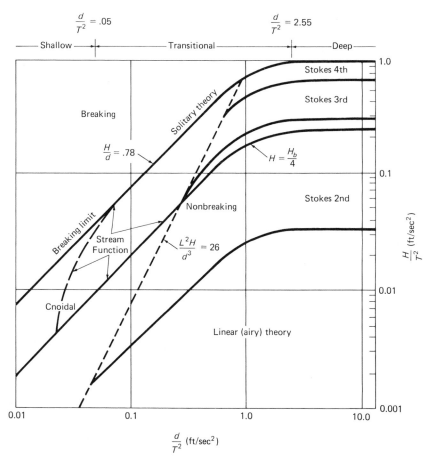

Fig. 3.38. Regions of validity of various wave theories. (After LeMehaute.[67])

considered; whereas where inertial forces predominate, linear theory is probably sufficient for deep water applications. Returning to Fig. 3.38, note that a curve representing the value of $HL^2/d^3 = 26$ has been plotted. This value, known as the Ursell parameter, has been proposed[68] as a line of demarcation to aid in the selection of theories.

The attempt, herein, is to describe the salient features and range of applicability of the subject theory rather than the mathematical derivation for which references are given. In general, only the two-dimensional case is considered.

Small Amplitude Wave Theory (The Airy Wave)

The solution for the wave equations can be found directly by linearizing the hydrodynamic equations; hence this method is a first order approximation. For this case, the velocity potential is given by:

$$\phi = \frac{H}{2}\frac{g}{\omega}\frac{\cosh k\,(\partial + y)}{\cosh kd}\sin(kx - \omega t) \qquad (3.33)$$

Equation (3.30) becomes:

$$\frac{\partial^2 \phi}{\partial t^2} + g\frac{\partial \phi}{\partial y} = 0$$

Solving the above with ϕ as given by equation (3.33) yields:

$$\omega^2 = gk \tanh kd$$

$$\text{or}$$

$$C^2 = g\omega \tanh kd \qquad (3.1)$$
$$\text{(already introduced)}$$

This basic wave equation was introduced and discussed in Section 3.1. If the effects of surface tension are to be considered, the term $(\sigma/\rho)\,k$ must be added to equation (3.1) above, where σ is the surface tension. Note that values of surface tension and other mechanical properties of seawater are given in the appendixes. This term is negligible for all practical applications to gravity waves. The water surface elevation for the Airy wave reduces to:

$$\eta = \frac{H}{2}\cos\theta \qquad (3.34)$$

and the horizontal water particle velocity is given by:

$$u = \pi\frac{H}{T}\frac{\cosh k\,(d + z)}{\sinh kd}\cos\theta \qquad (3.35)$$

which reduces to, for the wave crest mean water elevation position:

$$u = \pi\frac{H}{T} \qquad (3.36)$$

The group velocity (C_g), which is the average speed of the wave train corresponding to the rate of transmission of energy, is given in terms of the individual wave celerity as:

$$C_g = nC$$

where:

$$n = \frac{1}{2} \left[1 + \frac{2\,kd}{\sinh kd} \right] \tag{3.37}$$

which becomes equal to about one-half for deep water and approaches unity in shallow water. Waves traveling from deep to shallow water undergo a change in height, where $H/H_0 = K_s$ (K_s, the shoaling factor, was discussed in Section 3.3). Figure 3.39 illustrates various wave parameters as functions of relative water depth d/L_0 as calculated from the Airy wave theory. A more complete tabulation of various wave parameters as functions of d/L_0 can be found in Wiegel.[28] Because of the fundamental importance of linear, Airy wave theory, Table 3.9, which summarizes the important relations over deep,

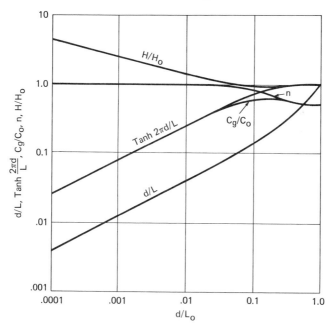

Fig. 3.39. Functions of d/L.

Table 3.9. Summary of linear wave equations.

Wave parameter	General expression (transitional water depths)	Deep water ($d/L > 1/4$)	Shallow water ($d/L < 1/20$)
Surface profile (η)	$\eta = \dfrac{H}{2} \cos(kx - \omega t) = \dfrac{H}{2} \cos \Theta$		
Celerity ($C = L/T$)	$C = \dfrac{g}{\omega} \tanh(kd)$	$= \dfrac{g}{\omega}$	$= \sqrt{gd}$
Length (L)	$L = \dfrac{gT}{\omega} \tanh(kd)$	$= \dfrac{gT}{\omega}$	$= T\sqrt{gd}$
Group velocity (C_g)	$C_g = nC = \dfrac{C}{2}\left[1 + \dfrac{4\pi d/L}{\sinh 4\pi d/L}\right]$	$= \dfrac{C}{2}$	$= \sqrt{gd}$
Horizontal particle velocity (u)	$u = \dfrac{\pi H}{T} \dfrac{\cosh(k(z+d))}{\sinh(kd)} \cos \Theta$	$= \dfrac{\pi H}{T} e^{kz} \cos \Theta$	$= \dfrac{H}{2}\sqrt{\dfrac{g}{d}} \cos \Theta$
Vertical particle velocity (W)	$W = \dfrac{\pi H}{T} \dfrac{\sinh(k(z+d))}{\sinh(kd)} \sin \Theta$	$= \dfrac{\pi H}{T} e^{kz} \sin \Theta$	$= \dfrac{\pi H}{T}\left(1 + \dfrac{z}{d}\right) \sin \Theta$
Horizontal particle acceleration (a_x)	$a_x = \dfrac{\pi H g}{L} \dfrac{\cosh(k(z+d))}{\cosh(kd)} \sin \Theta$	$= 2H\left(\dfrac{\pi}{T}\right)^2 e^{kz} \sin \Theta$	$= \dfrac{\pi H}{T}\sqrt{gd} \sin \Theta$
Vertical particle acceleration (a_z)	$a_z = \dfrac{\pi H g}{L} \dfrac{\sinh(k(z+d))}{\cosh(kd)} \cos \Theta$	$= -2H\left(\dfrac{\pi}{T}\right)^2 e^{kz} \cos \Theta$	$= -2H\left(\dfrac{\pi}{T}\right)^2\left(1 + \dfrac{z}{d}\right) \cos \Theta$
Horizontal particle displacement (ξ)	$\xi = -\dfrac{H}{2} \dfrac{\cosh(k(z+d))}{\sinh(kd)} \sin \Theta$	$= -\dfrac{H}{2} e^{kz} \sin \Theta$	$= -\dfrac{H}{2\omega}\sqrt{\dfrac{g}{d}} \sin \Theta$
Vertical particle displacement (ς)	$\varsigma = \dfrac{H}{2} \dfrac{\sinh(k(z+d))}{\sinh(kd)} \cos \Theta$	$= \dfrac{H}{2} e^{kz} \cos \Theta$	$= \dfrac{H}{2}\left(1 + \dfrac{z}{d}\right) \cos \Theta$
Pressure below surface (P)	$P = \rho g \eta \dfrac{\cosh(k(z+d))}{\cosh(kd)} - \rho g \varsigma$	$= \rho g \eta e^{kz} - \rho g \varsigma$	$= \rho g(\eta - z)$
Velocity potential (Φ)	$\Phi = \dfrac{HC}{2} \dfrac{\cosh(k(z+d))}{\sinh(kd)} \sin \Theta$	$= \dfrac{HC}{2} e^{kz} \sin \Theta$	$= \dfrac{Hg}{2\omega} \sin \Theta$

transitional, and shallow water, is included here. Actually the deep water relations can be applied in water depths as shallow as $d \geqslant L/4$ with only a 5% error. The effects of higher-order terms on the Airy wave length (L_A) are shown in Fig. 3.40, which presents a correction factor to wavelength due to wave steepness not accounted for by first-order theory.

Stokes Waves

Stokes[69] was the first person to develop equations for waves of "finite amplitude" by considering terms of higher than first order in solving the LaPlace equation. Stokes equations were later modified, and Skjelbreia and Hendrickson[70] prepared tables for practical application to help reduce errors in the tedious numerical computations. The Stokes waves, of successively higher order, give wave surface profiles that are steeper in the crests and flatter in the trough than those given by small amplitude wave theory, and which more closely resemble waves actually observed in the oceans. The surface profile is given by, for second order theory;

$$\eta = a \cos \theta + \frac{k}{2} a^2 \cos 2\theta$$

or for third order:

$$\eta = a \cos \theta + \frac{k}{2} a^2 \cos 2\theta + 3 \frac{k^2}{8} a^3 \cos 3\theta$$

These equations have been applied for terms up to fifth order. Another important aspect of the higher-order Stokes waves is that the particle orbits are not closed, as assumed by linear theory, and thus the phenomenon of mass transport is accounted for, such waves being known as progressive waves.

Stokes further determined that should the included angle between two tangents to the surface profile at the wave crest become less than 120°, the wave would become unstable (i.e., the particle velocity at the crest would exceed the celerity) and break. This corresponds closely with observations of deep water waves, but Michell[71] and later

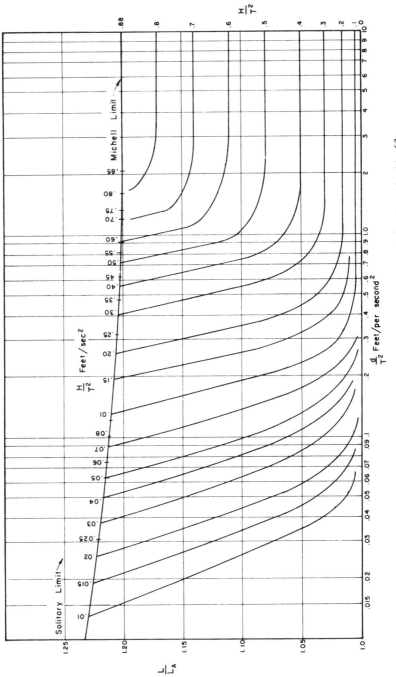

Fig. 3.40. Correction factor for wave length due to steepness. (After Bretschneider[63])

Havelock[72] demonstrated the theoretical limit for maximum steepness, in deep water to be:

$$\frac{H_0}{L_0} = .142 \approx \frac{1}{7} \qquad (3.28)$$
$$\text{(already introduced)}$$

The corresponding wavelength and celerity are given by:

$$L = 1.2\,L_0 \quad \text{and} \quad C = 1.2\,C_0$$

Stream Function Theory

Dean[73] first presented a numerical approximation to the solution of the hydrodynamic equations, which was later developed further by Dean[74] and by Monkmeyer.[75]   The stream function theory is a non-linear theory similar to the higher Stokes waves and with a relatively wide range of applicability. This theory is thought by many to better describe the actual wave phenomenon over a wider range of relative water depths than the other theories, and it has a broad range of application in determining wave forces on structures.

As with the other higher order theories, long tedious computations are required in evaluating all of the higher order terms; hence, the practical application of the theory utilizes tables and graphs, which have been presented by Dean.[76]

Cnoidal and Solitary Waves

In progressively shallower water the Stokian waves become less accurate at predicting the wave profile and particle velocities. The series expansion solution to the basic hydrodynamic equations must be carried out in terms of H/d instead of H/L. Although solutions have been developed, they are quite difficult to apply, and involve the use of elliptic functions. Masch has developed tables of cnoidal wave functions to expedite the use of this theory.[77] Tables of elliptic functions are also required. The term cnoidal waves in fact originates from the elliptic cosine function, which is abbreviated cn. Use of cnoidal theory involves the elliptic parameter (m), which always has a value between 0 and 1. When $m = 0$ the elliptic functions degenerate into sinusoidal functions and the wave becomes nearly Stokian; and when $m = 1$, the elliptic functions degenerate into hyperbolic functions. For this latter case, of $m = 1$, the wave is considered as a solitary wave. Cnoidal theory applies approximately over the range: $0.2 < d/L < .10$, and, therefore somewhat overlaps the regimes of the Stokian and solitary waves at the deeper and shallower water sites, respectively. Because this theory is applied with some difficulty and is not in routine engineering use, it will not be discussed further here. The limiting case, where $m = 1$, which is particularly useful in very shallow water, has been the subject of many engineering investigations 78–81 and will be reviewed briefly as follows.

In very shallow water wave crests become peaked and troughs flattened. The wave crest form becomes independent of its length and period and appears to be a solitary, translatory, wave with its entire crest above the still water level. The solitary wave has proved useful in many engineering applications such as the study of very large waves like tsunamis (see Section 5.5), and in determining wave properties near breaking in shallow water and for studying waves of maximum steepness in deep water. For a solitary wave of infinite length with its crest entirely above still water, the maximum height at breaking is given by:

$$H_b = .78d \tag{3.38}$$

and the corresponding celerity is:

$$C = \sqrt{2gH_b} \tag{3.39}$$

The water surface elevation is given by:

$$\eta = H \operatorname{sech}^2 \sqrt{\frac{3}{4} \frac{H}{d^3}} (x - Ct)$$

The horizontal water particle velocities are given by:

$$u = NC \frac{1 \pm \cos (Mz/d) \cosh (Mx/d)}{[\cos (Mz/d) + \cosh (Mx/d)]^2}$$

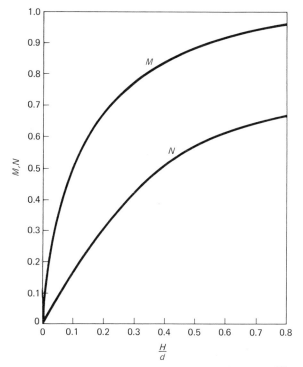

Fig. 3.41. Solitary wave functions M and N. (After Munk.[79])

The quantities M and N depend upon the ratio H/d and can be determined from Fig. 3.41 after the work of Munk.[79] The wave celerity can be approximated from the relation:

$$C = \sqrt{gd} \quad (1 - \frac{1}{2} H/d) \qquad (3.40)$$

An alternate simplified derivation for the celerity of a solitary wave front can be demonstrated using Bernoulli's equation as follows. Referring to Fig. 3.42, consider the steady flow case where the wave form shown remains stationary and water enters a velocity $V_1 = C$; the energy equation between points 1 and 2 is:

$$d_1 + \frac{V_1{}^2}{2g} = d_2 + \frac{V_2{}^2}{2g}$$

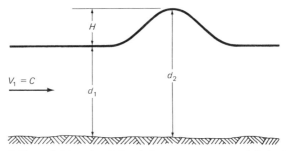

Fig. 3.42. Definition sketch for shallow water wave speed derivation.

The continuity equation gives:

$$d_1 V_1 = d_2 V_2$$

Substituting and rearranging:

$$\frac{V^2}{2g} = \frac{d_2 + d_1}{1 - (d_2/d_1)^2}$$

If $d_1 = d$, and $d_2 = d_1 + H$, and dropping terms of higher order:

$$C = \sqrt{gd}\left(1 + \frac{3}{2}\frac{H}{d}\right)^{1/2} \approx \sqrt{gd}\left(1 + \frac{3}{4}\frac{H}{d}\right)$$

Note that when height becomes small with respect to depth, the limit is:

$$C = \sqrt{gd} \tag{3.4}$$

(already introduced)

For extremely shallow water it is sufficiently accurate to take:

$$C = \sqrt{g(d + H)} \tag{3.41}$$

The Trochoidal Wave

The trochoidal wave theory was first presented by Gerstner.[82] Despite the fact that it has certain inaccuracies with respect to real waves and is not used for predicting kinematic properties, the trochoidal wave form is quite useful for illustrative purposes, and its simple geometry makes it convenient to work with. Naval Architects have traditionally used the $L/20$ trochoidal wave profile in calculating wave bending moments in ships. The trochoidal wave applies to oscillatory (i.e., water particles undergoing an orbital motion) waves of low amplitude. The theory assumes that the water particles undergo a circular motion with constant angular velocity as the wave form passes. Referring to Fig. 3.43, a surface particle orbit has a radius (r_0) of one-half the wave height, and the radii of particle orbits below the surface decrease exponentially with depth as given by:

$$r_0 = r\,e^{-kz}$$

where k is the wave number and z is the elevation measured from the still water level positive downward to the center of orbit. The wave form is generated by a circle of radius (R) rolling along a horizontal straight surface. The path described by a point on an inscribed circle of radius (r_0) is a trochoid, which gives a good description of a deep water, regular wave for steepnesses up to about $1/20$. The limiting case for $R = r_0$ is a cycloid, which has cusps much too pointed for real waves. The profiles formed by the decreasing radii of the orbits become flatter with depth and correspond to contours of equal pressure. It can be shown that, since the amount of fluid taken from the trough must equal the amount gained by the crests, the centers of orbit of the surface water particles are a distance of $r_0{}^2/2R$ above the still water level, which is consistent with the fact that more than one-half of the wave's height is above still water.

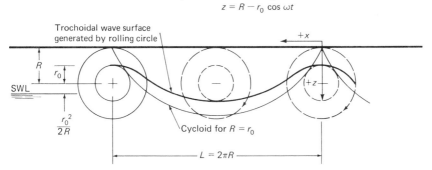

Parametric equations: $x = R\omega t + r_0 \sin \omega t$
$z = R - r_0 \cos \omega t$

Fig. 3.43. Trochoidal wave geometry.

As such a wave enters progressively shallower water, the particle orbits become more elliptical, and the velocity of propagation can be found by multiplying the deep water velocity (C) by $\sqrt{b/a}$, where b/a is the ratio of the vertical, semi-minor, to horizontal, semi-major, axis of the elliptical water particle motion. The parametric equations of motion are:

$$X = R(\omega t) + r_0 \sin(\omega t) \qquad (3.42)$$

and

$$Z = R - r_0 \cos(\omega t) \qquad (3.43)$$

Wave Energy

The total energy in a wave is composed of kinetic energy (E_K) due to the orbital motion of the water particles and of potential energy (E_p) due to the difference in elevation (trough to crest) of the particles. An equation for the kinetic energy per unit width of crest over one wavelength can be derived by integrating the squares of the horizontal and vertical components of the water particle velocities between the seabed and surface over one wavelength and multiplying by one-half the mass density; thus:

$$E_K = \frac{1}{2}\rho \int_0^L \int_{-d}^0 (u^2 + w^2)\,dx\,dz \qquad (3.44)$$

Inserting the expressions for u and w as derived from the velocity potential from linear theory, and carrying out the integration and algebra, the kinetic energy is found to be:

$$E_K = \frac{1}{16}\gamma H^2 L$$

The potential energy can be found by taking static moments about the still water level and integrating over the wavelength, and it is found to be:

$$E_P = \frac{1}{16}\gamma H^2 L$$

Thus the total energy over one wavelength is:

$$E = \frac{1}{8} \gamma H^2 L \qquad (3.45)$$

and the total energy per unit area of sea surface is:

$$E = \frac{\gamma}{8} H^2 \qquad (3.46)$$

The fact that wave energy is proportional to the square of the height is critical to the application of sea spectra for many types of engineering analysis.

The concept of wave power is also of engineering interest. Wave power is the energy transmitted per unit time propagated in the direction of wave travel. The relationship for wave power can be derived by considering the product of the force produced by the wave acting on a vertical plane normal to the direction of propagation times the particle velocities across this plane. The power can thus be shown to be:

$$P = \frac{n E}{T} \qquad (3.47)$$

where n is the value relating the group velocity to the wave celerity as given by linear theory, equation (3.37), and corresponds to the fraction of energy that is transmitted forward in a wave, E is the wave energy, and T is the wave period.

3.6 WAVE FORCES

General Considerations

The precise evaluation of the forces exerted upon an object by an ocean wave is extremely complicated because of many interacting phenomena. Among the most important of these factors are: the nonlinearity of the water particle displacements and kinematics, the variability of the wave profiles and forces, turbulence, the modification of the wave properties by the presence of the structure, and the possibility of dynamic effects such as vortex shedding and structure resonance. Wave force equations also represent an unsteady, nonuniform flow condition for which dynamic effects are nondeterministic. The calculation of wave forces first involves the selection of an appropriate wave theory to describe the particle kinematics and displacements for the given design wave condition, and, second, a choice of a semi-empirical coefficient or coefficients to reconcile the assumptions of theory with the findings of laboratory and field investigations. The attempt herein is to familiarize the reader with some of the various accepted methods of calculating wave forces, emphasizing practical applications and referencing important findings, rather than presenting rigorous theoretical justifications.

Waves impose various kinds and magnitudes of forces on structures, depending upon the structure type and configuration, and upon the wave characteristics (i.e., deep or shallow water and breaking, broken, or nonbreaking waves). A tentative classification of wave forces is presented in Fig. 3.44. Some of these conditions are illustrated in Fig. 3.45. Note in Fig. 3.44 that the various types of wave action relate to different structure types. The more important modes of interaction have been indicated in the figure, and the width of the arrows is intended to emphasize further the relative practical impor-tance of the particular wave–structure interaction. The nature of the predominating type of force is also indicated. Note that wave-induced motions give rise to accelerations and hence forces in floating bodies which can be further distinguished as to whether freely floating or fixed moored.

The emphasis of this section will be on contemporary civil engi-neering structures, such as piers and platforms, pipelines, caissons, walls and bulkheads, and so on, while only a brief overview will be given of forces on rubble mound structures such as breakwaters and jetties and of floating structures such as semi-submersibles, mobile breakwaters, and others. The subject of rubble mound structures is rather thoroughly discussed in various texts on traditional coastal engineering and shore protection. The subject of floating structures is becoming increasingly important, but the complexity of the inter-actions of waves and floating bodies is well treated in various naval architectural texts and would be too voluminous to treat in sufficient detail herein. As noted in the introduction (Chapter 1), floating structures are being used for many diverse applications outside the scope of traditional naval architecture, (e.g., floating plants, berths, dry docks, bridges, etc.). However, they are often located at sheltered

Type of Wave Action   Type of Structure

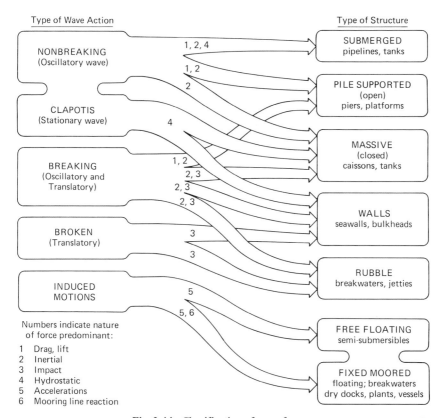

Fig. 3.44. Classification of wave forces.

sites where wave action is not significant. For these structures ocean tow from the construction site to final location is likely to be the worst environmental loading condition for which the structure must be designed.

The following discussion is organized around the predominant wave forces with regard to structure type. In general, breaking waves exert far greater loads than reflected or nonbreaking waves. Some structures may be subjected to both types, as noted in Section 3.4. Broken waves are generally a factor only on seawalls and bulkheads that are situated well beyond the breaking wave limit. As a practical matter, deep water can be considered as a depth corresponding to $d/T^2 \geqslant 2.5$ ft/sec^2, although the classical deep water equations can be applied to relative depths corresponding to $d/T^2 \geqslant 1.5$ ft/sec^2 without significant

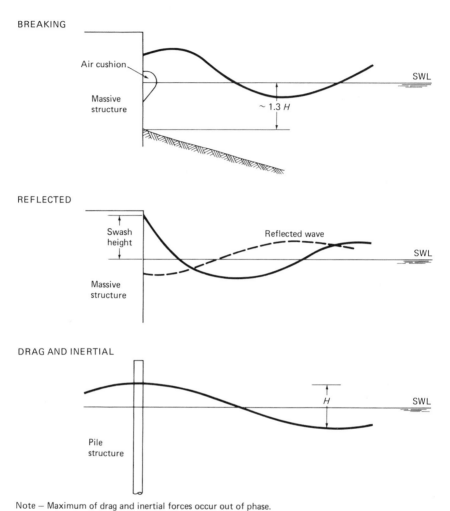

Note — Maximum of drag and inertial forces occur out of phase.

Fig. 3.45. Types of wave forces. (From Gaythwaite.[2])

error. Generally speaking, however, on the open ocean exposed to large storm waves deep water will apply to depths of around 200 to 300 feet or deeper.

It must also be remembered that even though some of the methods of wave force calculation may be conservative, and the choice of design wave parameters may also be conservative, a nominal safety factor must be applied in design. This factor may range from about

1.5 to greater than 2.0, depending upon the nature of wave action, structure type, probability of equaling or exceeding design wave conditions, confidence in calculations, and so on. The safety factor may also be given in terms of some "limit state" or "serviceability condition" as employed by some offshore codes.[47, 60, 83]

Modeling of structures to determine forces has been tried, but in general problems with scale effects, accuracy of measurements, and so forth, often render the results as questionable as those based on calculation. Plate and Nath[84] discuss the technique of modeling structures subjected to wind waves, and Apelt and MacKnight[85] present the results of model tests on a large caisson type offshore structure and compare these results to theoretical predictions. Full-scale measurements and prototype measurements are relatively scarce and sometimes proprietary. Some results are published,[86-90] which provide valuable information; however, such results are most valuable in correlating with theoretical calculations, as they are often not directly applicable to other, nonsimilar structures. Such results are difficult to obtain for and correlate with storm wave, design conditions and are difficult to correlate with other results and with theory because of the irregularity and directionality of real waves and the presence of other interacting phenomena such as ocean currents.

Pile Structures

Most of our knowledge regarding the forces exerted by waves on pile type structures is due to the offshore oil industry, and it is still a relatively new body of knowledge. Investigations of forces on pile structures began in earnest after World War II, and have been increasing steadily because of the deeper depths in which offshore platforms are being situated. In deep water platforms ($d > 300'$ ±) dynamic effects become all-important and dominate the design analysis. Such structures may have natural periods on the order of several seconds and hence be near resonance with potentially damaging waves. These effects will be discussed briefly later in this section; but most of our discussion will be devoted to what can be considered a routine static analysis of wave forces and moments, neglecting any dynamic amplification due to the structure's response. In most cases for shallower water pile-supported structures having natural periods on the order of 1 to 2 seconds or less, a static analysis and perhaps a check of the structure's fundamental period is sufficient, except, perhaps, in very shallow water with strong currents where vortex shedding phenomena and hence lateral vibrations are possible.

The following treatment assumes that the piles themselves have negligible effect on the wave properties. This assumption is mostly true when the ratio of pile diameter (D) to wavelength (L) is quite small, and when piles are sufficiently spaced to allow relatively uninterrupted passage of the wave. The pile diameter can be considered small when $D/L \leqslant .05$. Pile structures for which $D/L > .05$ are considered as large-diameter piles and will be discussed in the next subsection with massive structures.

The analysis of the force exerted on a single vertical cylindrical pile by a regular wave will be introduced, and then its application to pile groups and other pile geometries and orientations will be discussed. The following analysis was originally derived for cylindrically shaped piles, as their geometry is easier to work with, they present the same shape factor to any direction of wave approach, they present relatively low resistance to flow, and they are used almost universally in the construction of offshore structures.

In 1950, Morison et al.[91] introduced an equation for the analysis of wave forces on piles considering that the total force is due to an inertial force component arising from the water particle accelerations, and a drag component due to friction and boundary layer effects. This method is still employed for static pile load analysis. A fixed object in an accelerating fluid behaves as though it has an increased mass because of a virtual layer of entrained fluid about the object. Newton's second law for an added mass (M') then can be written: $F = (M + M') \, du/dt$. Morison and his colleagues demonstrated this to be, for a cylindrical pile of diameter (D):

$$F_i = \rho \, C_m \, \frac{\pi D^2}{4} \, \frac{du}{dt} \qquad (3.48)$$

where C_m is an experimentally determined mass coefficient, which has a theoretical maximum value of 2.0. This equation can be rearranged, introducing the wave height (H), to yield:

$$F_{im} = \frac{1}{2} \, \rho \, C_m \, K_{im} \, D^2 H \qquad (3.49)$$

Where F_{im} is the maximum value of the inertial force, and K_{im}, called the inertial force factor, represents an integration of the horizontal water particle accelerations from the free surface to the seabed as given by:

$$K_{im} = \frac{\pi}{2H} \int_0^{\eta_s} \frac{du}{dt} dz \qquad (3.50)$$

where η_s is the free surface elevation such that F_{im} is maximum.

The drag portion of the total force results from friction or boundary layer shear and from pressure or form drag due to the separation of flow at the downstream side of the cylinder. Boundary layer theory (covered in any fluid mechanics text) assumes that an infinitely thin layer of fluid adheres to the side of the cylinder, remaining stationary, and that a velocity distribution is thus set up within layers of fluid which increases exponentially with distance from the cylinder surface. The drag force arises as a result of internal shear between adjacent layers of fluid and as a result of the boundary layer breaking away from the after side of the cylinder (which occurs at certain relative flow velocities dictated by the Reynolds Number, N_R, causing a turbulent wake of lower pressure and a net force in the direction of fluid motion). This force is given by the well-known drag force equation introduced in Section 2.3:

$$F_d = \frac{1}{2} \rho C_D V^2 A \qquad (3.51)$$

Morison and his fellow workers showed this could be represented for the maximum drag force on a pile as:

$$F_{dm} = \frac{1}{2} \rho C_D K_{dm} DH^2 \qquad (3.52)$$

where C_D is an experimentally determined drag coefficient and K_{dm} represents an integration of water particle velocities between the free surface and seabed as given by:

$$K_{dm} = \frac{1}{H^2} \int_0^{\eta_c} u|u|dz \qquad (3.53)$$

where η_c is the surface elevation at the crest position.

The total maximum force on a vertical cylindrical pile is then, according to Morison:

$$F_{TM} = \frac{1}{2} \rho \, C_m \, K_{im} \, D^2 H \; \overrightarrow{+} \; \frac{1}{2} \rho \, C_D K_{dm} DH^2 \qquad (3.54)$$

The evaluation of the total maximum force is then reduced to one of determining the drag and inertial coefficients and factors. When currents are present, the steady-state current velocity must be added to the wave particle velocities vectorially in order to solve for the total combined force. The evaluation of particle velocities and accelerations depends upon the wave theory applied. Figures 3.46 and 3.47, based upon the work of Reid and Bretschnieder,[92] can be used to determine K_{dm} and K_{im} for the given wave period parameters of relative depth, d/T^2, and relative steepness, H/T^2. The curves in these figures form an envelope of values limited by Airy wave and solitary wave theories corresponding to waves of zero and maximum steepness, respectively.

Other methods of computing particle kinematics for wave force are available. Skjelbria et al.[93] prepared tables that can be used to

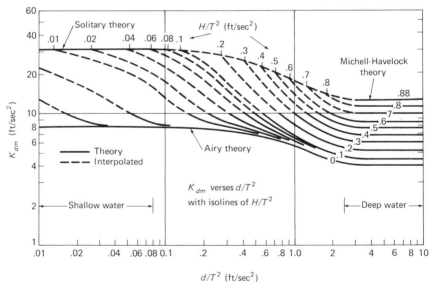

Fig. 3.46. K_{dm} vs. d/T^2. (After Reid and Bretschneider.[92])

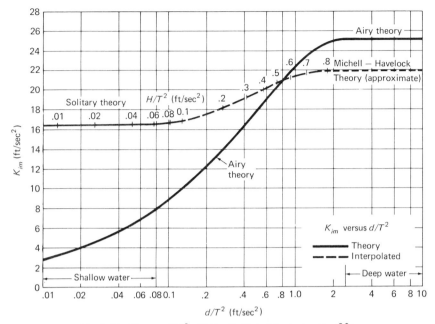

Fig. 3.47. K_{im} vs. d/T^2. (After Reid and Bretschneider.[92])

solve for K_{im} and K_{dm}, based on Stokes's fifth order theory, and Dean[76, 94] presented tables and graphs based on his numerical stream function theory, which is probably more accurate over a wider range of relative depths and steepnesses than any other single theory.

Because the maximums of F_{dm} and F_{im} occur approximately $90°$ out of phase (i.e., F_{dm} occurs at the crest position where velocities are greatest, and F_{im} occurs close to one-quarter of the wavelength ahead of the crest where the accelerations are greatest), the total maximum force, F_{Tm}, is a vector addition and is also dependent upon the period parameters, H/T^2 and d/T^2. Figure 3.48 can be used to obtain F_{Tm} from the ratio of F_{im}/F_{dm}.

The total moment about the mudline is of critical importance in structural design. The relative lever arms S_d and S_i above the bottom for the drag and inertial forces can be determined from Figs. 3.49 and 3.50. The total overturning moments about the bottom can be obtained from:

$$M_{dm} = S_d \, F_{dm} \quad \text{and} \quad M_{im} = S_i \, F_{im}$$

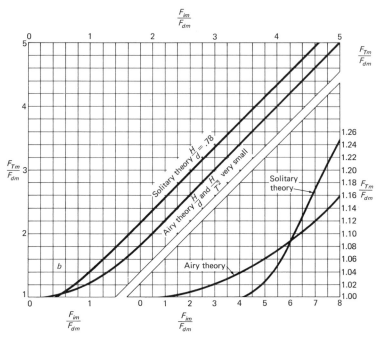

Fig. 3.48. F_m/F_{DM} vs. F_{im}/F_{Dm} for the breaking solitary wave and the airy wave. (After Reid and Bretschneider.[92])

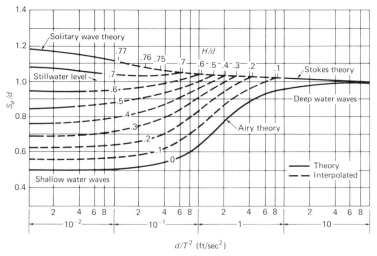

Fig. 3.49. Relative lever arm measured from bottom vs. relative depth corresponding to maximum drag force. (After Reid and Bretschneider.[92])

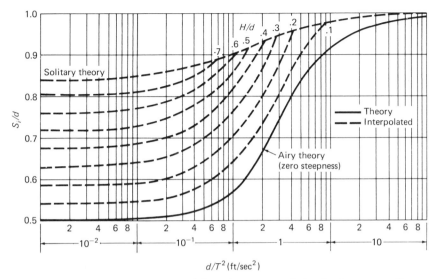

Fig. 3.50. Relative lever arm measured from bottom vs. relative depth corresponding to maximum inertial force. (After Reid and Bretschneider.[92])

where the ratio M_{im}/M_{dm} can be substituted in Fig. 3.48 to obtain the maximum total moment, M_{Tm}. As a matter of practical design, however, some length may be added to S_d and S_i to account for looseness of bottom material and distance to assumed location of pile fixity when computing moments for structural design of the pile.

The use of the curves in Figs. 3.46 through 3.50 can best be illustrated through an example problem. By working through such hypothetical problems and varying different input parameters each time, the reader will begin to acquire a feel for the relative effects of wave height and steepness, water depth, and so on, upon the forces and moments produced. We shall assume for our example problem that it is required to find the maximum total force and moment about the sea bottom exerted upon a single 36-inch-diameter vertical cylindrical pile situated in 60 feet of water, by an incident wave having a height of 20 feet and a period of 8 seconds. This example problem is illustrated in Fig. 3.51. Note that the water depth is some design water level (D.W.L.) that accounts for storm tide (see Section 5.2) as well as the astronomical tide, and it may be required in actual practice to check forces for various design water depths. Before the forces and moments are calculated, the maximum wave crest eleva-

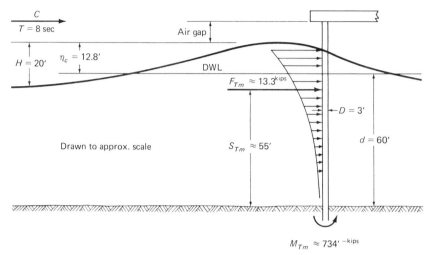

Fig. 3.51. Example wave force calculation on single vertical pile.

tion should be checked to be sure there is adequate under-deck clearance below the superstructure. Usually a safety margin called an air gap is added to the height of fixed platforms to be sure that no bouyant uplift or direct wave forces act on the deck structure itself. Before beginning the force and moment calculations, we should also check to be sure that the theory and application of the Morison equations apply by checking the pile diameter to wavelength criteria. All of the curves presented herein are based upon the period parameters d/T^2 and/or H/T^2 in foot, second units. For our example problem, then:

$$d/T^2 = 60 \text{ ft} \div (8 \text{ sec})^2 = .94 \text{ ft/sec}^2$$

and

$$H/T^2 = 20 \text{ ft} \div (8 \text{ sec})^2 = .31 \text{ ft/sec}^2$$

The relative depth parameter ($d/T^2 = .94$) indicates we are in a transitional water depth where neither deep water (Airy theory) nor shallow water (solitary theory) relations apply exactly. From Fig. 3.36, presented earlier, we find the nondimensional crest elevation to wave height ratio, $\eta_c/H = .64$; thus the height of the wave crest above the still water level will be: $\eta_c = .64 \times 20 \text{ ft} = 12.8 \text{ ft}$. The deep water wavelength as given by equation (3.3) is: $L_0 = 5.12T^2 = 5.12 (8 \text{ sec})^2 = 328 \text{ ft}$. This wavelength will be altered by the effects of relative water depth and relative wave steepness. The correction factor of wavelength for the effect of water depth for linear (Airy) wave theory can be obtained from Fig. 3.52, where L_A is the Airy wavelength given in terms of L_A/L_0 = .88, for our example, which yields L_A = .88 X 328' = 289 ft. This wavelength must in turn be corrected for the effects of steepness not accounted for by linear theory, as given earlier in Fig. 3.40; from this figure the actual wavelength (L) will be 1.04 X 289' = 301 ft. The ratio of pile diameter to wavelength is then 3 ft ÷ 300 ft = .01, which is less than .05; so the Morison equation can be applied. Entering Figs. 3.46 and 3.47 to find the maximum drag and inertial force factors gives K_{dm} = 10.6 ft/sec^2 and K_{im} = 22.4 ft/sec^2. Some interpolation between curves is involved to cover values not plotted. These values can now be substituted into equations (3.52) and (3.49), respectively. If we also select drag and inertial coefficients of C_D = 1.0 and C_M = 1.5 as being representative, then we can solve for the maximum drag and inertial forces as follows. From equation (3.52), the maximum drag force is:

$$F_{dm} = \frac{1}{2} \rho \, C_D K_D H^2 D = \frac{1}{2} \, \frac{64 \text{ pcf}}{32.2 \text{ fps}^2} \, (1.0) \, (10.6 \text{ fps}^2)$$
$$(20 \text{ ft})^2 \, (3 \text{ ft}) = 12,641 \text{ lb}$$

Similarly, from equation (3.49), the maximum inertial force is:

$$F_{im} = \frac{1}{2} \rho \, C_m \, K_m D^2 H = \frac{1}{2} \, \frac{64 \text{ pcf}}{32.2 \text{ fps}^2} \, (1.5) \, (22.4 \text{ fps}^2)$$
$$(3 \text{ ft})^2 \, (20 \text{ ft}) = 6,010 \text{ lb}$$

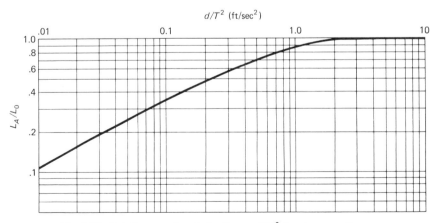

Fig. 3.52. Relative wave length L_A/L_O vs. d/T^2 for airy wave theory.

These forces occur out of phase and, therefore, must be added vectorially. Figure 3.48 can be used to obtain the maximum total force (F_{Tm}) in terms of the ratio of F_{im}/F_{dm} = 6,010 lb ÷ 12,641 lb = .47; from the figure we get F_{Tm}/F_{dm} = 1.05 (some interpolation is required here), from which we can calculate F_{Tm} = 1.05 × 12,641 lb = 13,273 lb, which is the maximum total combined force acting on the pile.

Figures 3.49 and 3.50 can be used similarly to find the height of the drag and inertial force components above the seabed. We must first calculate the ratio H/d = 20 ft/60 ft = .33, to enter these figures and find the relative lever arms S_d/d = .94 and S_i/d = .78, from which we can calculate the moment of the wave force about the bottom as:

$$M_{DM} = .94 \times 60' \times 12,641 \text{ lb} = 712,952 \text{ ft lb}$$

and

$$M_{iM} = .78 \times 60' \times 6,010 \text{ lb} = 281,288 \text{ ft lb}$$

The maximum total moment can be found by substituting the moments for forces in Fig. 3.48 or M_{iM}/M_{dM} = .39, which gives from the figure M_{Tm}/M_{dM} = 1.03, from which we can calculate: M_{Tm} = 1.03 × 712,952 = 734,341 ft lb. Note that in practice several feet may be added to the moment lever arm to account for the depth to fixity of the pile. It should further be borne in mind that the use of these figures is sufficiently accurate for preliminary estimates of wave forces and moments and as an alternate checking method. In addition, because these curves are relatively convenient and easy to use, they can also be used instructionally to give one a feel for the effects of varying parameters, such as water depth, wave height and period, and so on. For final design, however, there are commercial computer programs available as well as more rigorous techniques for finding wave forces as given in the various references cited herein.

There have been many investigations within the last three decades to determine the values of C_D and C_M. Some experimental results have been summarized in references 34, 46, and 86. The selection of a final drag or inertial coefficient ultimately becomes one of the designer's subjective judgments. These coefficients vary with pile roughness, degree of fouling, aspect ratio, phase position, relative flow velocity, and so forth. In particular, the drag coefficient varies with the Reynolds number of flow, N_R, as illustrated in Fig. 2.9. For circular cylinders three distinct flow zones can be defined, that of subcritical flow and relatively high but constant C_D, a transitional zone of widely varying C_D, and a supercritical zone of more turbulent flow, where C_D takes on its minimum values. The U. S. Army Coastal Engineering Research Center[22] recommends values of C_D ranging from C_D = 0.7 for N_R = 5 X 10^5 (supercritical) to C_D =1.2 for N_R = 2 X 10^5 (subcritical). Similarly for the mass coefficient C_m , reference 22 gives ranges of C_m = 1.5 for N_R = 5 X 10^5, to C_m = 2.0 for N_R = 2.5 X 10^5. Reid and Bretschneider[92] recommended mean values of C_D = 0.5 and C_M = 1.5. For rough cylinders Goda[95] recommended values of C_D = 1.0 and C_M = 2.0 where the particle velocities and accelerations have been accurately calculated. Laboratory determinations of C_D and C_m are made difficult because of the many interacting phenomena and theoretical considerations. C_D and C_m do not remain constant along the wavelength. Experimentally determined coefficients range from about 0.4 to 1.8 for C_D and from about 1.0 to 2.3 for C_m. The reader is referred to the literature for further discussion.

Little information is available for pile shapes other than cylinders. Morison et al.[96] conducted tests on flat plates, and H-pile sections, as well as cylinders, and determined that the wave forces were smallest for cylinders, increasing about 25% for flat plates of the same projected width, and increasing from 42% to 158% for H-sections perpendicular to the flow, and to 122% to 258% when oriented at 45° to the flow.

The problem of calculating forces on nonvertical, oblique piles is more complex. In general, the approach is to obtain the force per unit length of a fictitious vertical pile situated at various positions along the locus of the inclined pile. An example of this approach is given in reference 22.

The total wave force on a group of piles can be obtained by superposition of the wave profile and by summing the forces on the individual piles. Care must be taken, however, to account for local proximity effects. For the case of piles in line with the direction of flow, it is probably safe to neglect proximity effects in wave force calculations when the clear spacing between piles is greater than about three times the pile diameter.[22] Dean and Harleman[34] have summarized some test results and note that for closely spaced piles in line with the flow, the downstream pile will likely receive some sheltering effect due to the turbulent wake of the upstream pile, which effectively reduces the drag force on the downstream pile. Test results are far from conclusive, however, and it is probably wise not to consider that any sheltering will take place in design force calculations. In fact, for certain relative pile spacings and orientations the effective total wave force may be considerably increased. Quinn[97] notes that tests with piles arranged in rows indicate the center pile in a row of three piles parallel to the wave crest receives a moment of up to 2.4 times the moment for a single pile when the gap between piles is only $.5D$. This increase in moment becomes negligible when the gap is increased to $1.5D$ or greater. For a similar arrangement of three piles perpendicular to the wave crest a sheltering effect allows for an average 30% reduction in moment on the center pile, which probably should be neglected in design. Quinn also provides an example of wave force calculations for a hypothetical offshore platform in several water depths.

If it is assumed that C_D and C_m remain relatively constant over the wavelength, then the force along the length of the wave can be reasonably approximated from:

$$F_d = F_{dm} \cos \theta \ \left| \cos \theta \right| \qquad (3.55)$$

and

$$F_i = F_{im} \sin \theta \qquad (3.56)$$

where θ is the phase angle = $(kx - \omega t)$ and $\theta = 0$ at the crest position. Figure 3.53 shows the variation in F_i, F_d, and F_t for a hypothetical design wave and platform structure. In the first position, the wave is positioned so as to give the maximum force on the first pile, and in the second position the wave is located so as to give the maximum total horizontal force and moment on the whole structure. By plotting the variation in total wave force over a wavelength as obtained from summing equations (3.55) and (3.56), the total maximum force on a multi-pile structure, having piles of the same diameter spaced sufficiently far apart, can be found by successively positioning the total force envelope to find the position of maximum total force on the group.

There is little information available on the problem of waves breaking on pile structures. In deep and transitional water, the method

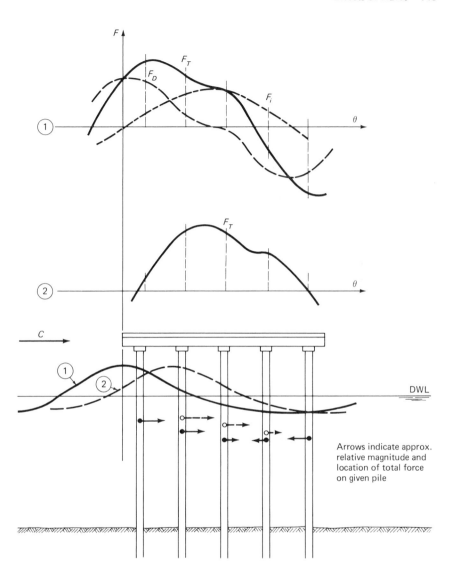

Arrows indicate approx. relative magnitude and location of total force on given pile

① Location of wave profile for maximum force on first pile in group

② Location of wave profile for maximum force on group — Total force = sum of ordinates of F_T at each pile.

Fig. 3.53. Illustration of total force on pile structure.

just discussed can be used, considering the breaking wave limit in obtaining coefficients and force factors. In very shallow water, the inertial force becomes very small with respect to the drag force, and the total maximum force can be considered as consisting of the drag component only, as given by:

$$F_{m_b} = \frac{1}{2} \gamma \, C_b DH_b{}^2 \qquad (3.57)$$

Hall[98] conducted small-scale experiments, at $N_R = 5 \times 10^4$, on beaches of slope $m = 0.1$ and found the coefficient C_b to have a mean value of about 1.5 and a maximum value of 3.0. Using this equation, the corresponding moment about the bottom is approximately equal to the product of equation (3.57) and the breaker height H_b, or:

$$M_{m_b} = F_{m_b} \, H_b \qquad (3.58)$$

Large-scale experiments by Ross[99] seem to support the lower values of C_b. Reference 22 recommends a value of $C_b = 1.75$.

Forces on piles due to broken waves, such as on piers situated on beaches extending into the surf zone, are generally small compared to the breaking wave forces and will not be discussed further here.

At shallow water sites, and especially those where strong steady state currents are present, transverse "lift" forces may occur as a result of the phenomenon of vortex shedding. At certain flow speeds, as dictated by N_R, a periodic shedding of turbulent vortices from alternate sides of the pile may occur, giving rise to transverse lift forces. The frequency of vortex shedding is given by the Strouhal number (N_S), introduced in Section 2.4 as:

$$N_S = \frac{Df}{V}$$

where f is the frequency and V is the flow velocity. The theory for vortex shedding is rather well developed for the case of steady non-uniform flow but not so well developed for the unsteady periodic flow of waves. The steady flow case is discussed to some extent in connection with wind forces in Section 2.4, and values of N_S for various shapes are given in Table 2.4. Various eddy-shedding regimes can be defined dependent upon the Keulegan–Carpenter[100] number

(K_c) as given by $K_c = \bar{U}_{max} T/D$ where $\bar{U}_{max}$ is the maximum horizontal particle velocity averaged over the water depth, and T and D are the wave period and pile diameter, respectively. When K_c has a value less than approximately 3.0, few or no eddies are formed, and transverse lift forces are effectively zero.

For values of $K_c > 3.0$ the transverse lift force increases relatively uniformly for a given value of H/T^2 and may equal about 50% of the drag force at $K_c = 10$. Bidde[101] correlated values of C_L/C_D with K_c where C_L is the lift force coefficient. The magnitude of the lift force can be expressed similarly to the drag force (equation 3.51) as:

$$F_L = \frac{1}{2} \rho C_L V^2 A \qquad (3.59)$$

where A is the projected area normal to the flow. This equation applies to the steady flow case only. The U. S. Army Shore Protection Manual (SPM)[22] presents a simple formula for estimating lift forces for the case of periodic flow. The maximum lift force is proportional to the square of the horizontal water particle velocity. Although the magnitude of this transverse force may be low, if the frequency of vortex shedding is near to one of the structure's natural modes, then resonant amplification of the forces and/or noisome vibrations may result.

In deep water, relatively tall, slender platform type structures may have periods long enough to coincide with the periods of large waves or with smaller waves possessing significant spectral energy at or near some resonant frequency. Nolan and Honsinger[42] note that the tragic loss of Texas Tower #4 off the coast of New Jersey in January 1961 was primarily due to resonant coupling of regular wave trains with heights considerably less than the design wave height. The tower's fundamental period had apparently been increased because of the deterioration of some cross bracing. The authors further point out that actual measurements of the tower's deflections prior to the failure gave platform displacements that were approximately 3 inches greater for 10- and 11-foot waves than for 30-foot waves! It is clear that the highest wave then may not be the most critical in structure design and that even though the structure's fundamental period may be well below that of the design wave (which has been more or less standard practice), the shorter periods of smaller waves may be critical to the structure's design. Nolan and Honsinger go on to present dimensionless curves of drag force multipliers and describe a procedure for estimating the dynamic amplification of wave forces. Their paper provides an interesting introduction to platform dynamics.

Structures may oscillate in various modes (e.g., flexural, torsional, coupled, etc.), which may be affected by the wavelength relative to the pile spacing, as shown schematically in Fig. 3.54. When the wavelength is equal to the pile spacing, all piles realize the same magnitude and direction of force simultaneously for a regular wave train. Nath and Harleman[102] reported on the effects of support spacing and wave direction on the dynamics of a four-legged tower model and presented an equation for optimum leg spacing for deep water platforms.

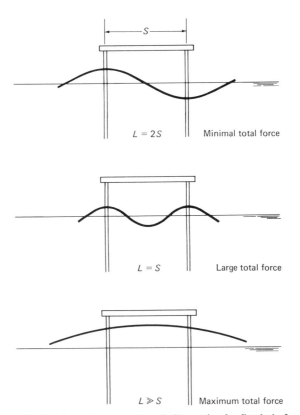

Fig. 3.54. Relation of wavelength and pile spacing for fixed platform.

Structural dynamics are beyond the scope of this book, although for most deep water structures dynamic analysis governs the design. Detailed dynamic analysis requires the use of high-speed computers. The reader is referred to the work of Brannon et al.,[103] who studied and reviewed dynamic design considerations; Malhotra and Penzien,[104] who employed spectral techniques to study platform dynamic response; and Wirsching and Prasthofer,[43] who presented a simplified method of preliminary dynamic assessment of deep water platforms, which provides a good introduction to the subject of platform dynamics. Further references dealing with this subject can be found in the bibliographies of the references cited above.

Large-Diameter Piles and Massive Structures

When the diameter of a pile is not negligible compared to wavelength, approximated by $D/L > .20$, the presence of the pile will modify the wave properties, causing reflection and diffraction and other local effects. Note that in Morison's equation the drag force varies directly with D, but the inertial force varies with D^2. Therefore, for large-diameter piles the drag force becomes small and the force on the pile, or object, is due primarily to inertial effects. MacCamy and Fuchs[105] investigated the interaction of a pile in a train of regular waves based on acoustical analogies, out of which they propounded the so-called diffraction theory. Their work is based upon linear theory and assumes a vertical circular cylinder continuous to the bottom in water of finite depth, and assumes that the wave is reflected from the pile vertical surface without loss of energy. Because this method is somewhat complicated, and its application is less common, it will not be presented herein. Those wishing a more detailed treatment of this analysis should refer to the original work by MacCamy and Fuchs or to references 28 and 46.

Diffraction theory can be applied with caution to various massive ocean structures such as caissons, gravity structures, large-diameter bases of lighthouses, and so on. Mogridge and Jamieson[106] presented a method of determining wave forces on square caissons based upon the linear diffraction theory of MacCamy and Fuchs. An approximate method for determining the maximum inertial force, moment, and corresponding phase angle is given by Mogridge and Jamieson, and the solutions can be found directly by calculation with values obtained from only three nondimensional graphs. Mogridge and Jamieson indicate this method has been found satisfactory for relative caisson sizes of b/L from .09 to .40, where b is the caisson width, and for relative water depths and wave steepnesses of d/L from .09 to .79 and H/L up to .09. The above ranges of applicability are independent of the orientation to the wave direction. For b/L and d/L less than .09, the method cannot be used except for waves of very low steepness, as viscous drag forces and nonlinearity of the waves become important. Other investigations of wave forces on large bodies of engineering significance can be found in references 85, 107, 108, and 109.

Submerged Objects, Pipelines, and Cables

For the case of large submerged objects, the diffraction theory of MacCamy and Fuchs can be applied, considering the appropriate boundary conditions. This is particularly so for small values of H/D, in consideration of the fact that particle orbits are proportional to wave height, and when this dimension is less than about half the structure dimension, there is little boundary layer development with consequent flow separation and resultant drag forces. Most of the references cited for massive structures contain information that can be applied to submerged structures as well. Chakrabarti[110] analyzed wave forces on submerged objects of symmetry that are an order of magnitude smaller than the wavelength, and that are confined near enough to the bottom that free surface effects are negligible. Vongvisessomjai and Silvester[111] summarized various experimental and theoretical works on submerged objects and looked at the ranges of applicability of the various methods of obtaining wave forces on submerged objects with regard to relative size and distance from bottom. The reader is referred to the literature for a more detailed treatment of wave forces on submerged structures.

For pipelines resting on the bottom, lift forces may become significant, especially where strong steady-state currents exist. The lift force, as given by equation (3.59), becomes small when the pipeline is elevated more than 0.5 pipe diameter above the bottom. The lift force effectively diminishes the vertical normal force, resulting in less resistance to sliding sidewise on the bottom. Pipelines are particularly subject to scouring and other problems beyond the scope of this book. A few selected papers on the hydrodynamics of underwater pipelines are included in the references.[112-114]

Cables, including flexible wire rope, fiber line, and chain, are frequently used to moor or tow various floating structures and buoys. They are subject to drag and inertial forces, and various dynamic effects including interaction with waves, currents, and moored body motions and harmonic oscillations generally known as strumming. Cable motions are very complex; finite element techniques are usually employed, which are beyond the scope of this book. Berteaux[115] provides an excellent introduction to mooring line loads and cable dynamics. Techniques for analyzing mooring line dynamics have been presented by Wilson and Garbaccio[116] and by Chang and Pilkey.[117]

Walls and Bulkheads

Rigid, vertical, or nearly vertical, walls such as concrete seawalls, steel or timber sheet pile bulkheads, quay walls, and vertical wall breakwaters are common to waterfront construction, and in many cases are subject to wave attack. If the water depth at the toe of the structure is deep enough, approximately 1.5 to 2 times the incident wave height, the wave will be reflected and form a standing wave pattern known as a clapotis. Depending upon the stage of tide and height of wave, the same structure may also be subjected to the greater forces exerted by a wave breaking upon it; or if the structure is situated at the head of a beach, for example, it may be subject to the impact of a traveling surge of water from a broken wave. The approaches to evaluating these three classes of wave forces are somewhat different, so they will be discussed individually in the following paragraphs.

Figure 3.55 illustrates the geometry and pressure distributions formed by a nonbreaking, clapotis wave. Assuming no energy losses upon reflection the sum of the reflected and incident wave heights is equal to $2H$, where H is the original deep water height. The resulting pressure on the wall consists of a hydrostatic and a dynamic component as given by:

$$P = -\gamma z \pm \frac{\gamma H \cosh k (d + z)}{\cosh kd} \cos kx \cos \omega t \qquad (3.60)$$

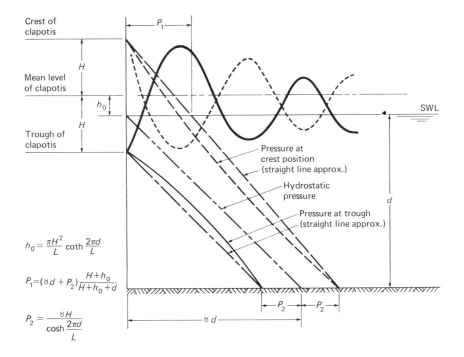

Crest of clapotis

H

Mean level of clapotis

h_0

SWL

H

Trough of clapotis

P_1

Pressure at crest position (straight line approx.)

Hydrostatic pressure

Pressure at trough (straight line approx.)

d

$$h_0 = \frac{\pi H^2}{L} \coth \frac{2\pi d}{L}$$

$$P_1 = (\delta d + P_2) \frac{H + h_0}{H + h_0 + d}$$

$$P_2 = \frac{\delta H}{\cosh \frac{2\pi d}{L}}$$

P_2 P_2

δd

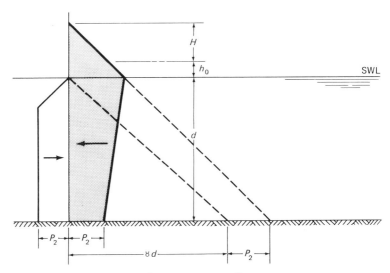

H

h_0

SWL

d

P_2 P_2

δd

P_2

Fig. 3.55. Standing wave pressure diagram.

The dynamic pressure at the toe of the wall fluctuates between plus and minus.

$$\pm \ \frac{\gamma H}{\cosh kd}$$

Finite amplitude effects cause the mean water level at the structure to rise by an amount given by:

$$\triangle z \ = \ \frac{\pi H^2}{L} \ \coth kd \qquad (3.61)$$

For the case of overtopping, the calculation of forces can proceed assuming the wall was extended above the wave runup and using a truncated pressure diagram. The problem of run-up and overtopping is discussed in the next section. Note that the minimum pressure corresponding to the wave trough position may be of critical importance, as some walls have failed outward because of active soil pressure or hydrostatic pressure behind the wall.

Several methods have been proposed for the calculation of standing wave pressures on vertical walls. Hudson[118] has summarized and compared the various theories and experimental results for the case of nonbreaking wave pressures on walls. The methods of Sainflou[119] and of Molitor[120] are perhaps the most commonly employed in engineering design. Sainflou's method is probably the most useful and shows good agreement with experimental results in a range of water depths around d/L = 0.1 to .02. For very shallow water, however, the method should not be used. Quinn[97] compares the Sainflou and Molitor methods in an example breakwater design problem. Minikin[121] suggests an effective pressure height of $1.66H$ for design problems, which should be a useful rule of thumb, at least for preliminary estimates. The SPM[22] and Nagai (in Bruun)[122] provide graphs based upon modified theoretical considerations and experimental data that cover an essentially complete range of relative water depths and can be used to solve almost any standing wave pressure problem. For walls with sloping faces, the maximum steepness of waves that will be fully reflected is given by:

$$\left(\frac{H}{L} \right)_{\text{max}} = \frac{\cos^2 \beta}{\pi} \ \sqrt{1 - \frac{2\beta}{\pi}} \qquad (3.62)$$

where β is the angle of inclination from the vertical.

Waves breaking on a structure cause instantaneous shock pressures of short duration, which are of considerably greater magnitude than reflected wave pressures. Because wave impact pressures are highly variable, all design methods are based upon empirical relations. The theoretical upper limit to the value of this shock pressure would be that of a full water-hammer (fluid compression wave) effect. This effect probably can never be fully realized in real waves because of the presence of an air cushion under the curl of the breaking wave crest. Bagnold[123] found that the presence of the air trapped under the curl of a plunging type breaking wave was responsible for the higher impact pressures. The pressure varies with the thickness of this air cushion and with the point of breaking at which the wave strikes the wall. Impact pressures have been found to correspond statistically to the heights of the incident waves. Denny[124] found, for the most frequently occurring maximum instantaneous pressures, that:

$$P_{max} = 28\gamma H$$

and for the extreme shock pressures observed:

$$P_{max} = 100\gamma H$$

These relations were determined from model experiments where the extreme pressures were on the order of duration of .01 second.

Figure 3.55 shows the geometry and pressure distribution assumed for design purposes of a wave breaking on a vertical wall. Minikin[121] developed the following formula for the maximum dynamic pressure for the conditions shown in Fig. 3.56:

$$P_{max} = 101\gamma \frac{H_b}{L_{d_1}} \frac{d_s}{d_1} (d_s + d_1) \qquad (3.63)$$

Where H_b is the breaker height, L_{d_1} is the wavelength in water depth d_1, d_1 is the depth one wavelength out from the face of the wall, and d_s is the water depth at the toe of the wall. The value of H_b and L_{d_1} should be based upon the deep water wave characteristics that have been corrected for the effects of shoaling and refraction, and so on, as discussed in Section 3.3. The peak pressure occurs at the still water

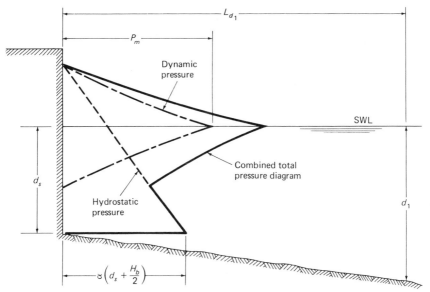

Fig. 3.56. Breaking wave pressure diagram for Minikin's formula.

level, and the total dynamic force must be obtained from the area within the pressure distribution, which is given by:

$$F_m = \frac{P_m H_b}{3} \qquad (3.64)$$

The corresponding overturning moment $M_m = F_m d_s$. The dynamic pressure component must be added to the hydrostatic pressure to get the total force on the wall. Minikin's formula was originally derived for composite breakwaters consisting of a concrete superstructure founded upon a rubble stone substructure. The formula may give very high values of dynamic pressure, but at the present time it is probably the most useful one for practical application.

In addition to impact and momentary hydrostatic pressure, breaking waves cause deflected vertical forces that tend to shear any projections off the face of the wall or dislodge material at the base of the wall. There is also a suction due to back draught, and in addition water or air trapped in the crevices of a porous wall may progressively break down the wall because of a repeated cycling of water hammer or air compression effect.

The dynamic pressure exerted by the turbulent uprush of a broken wave is difficult to evaluate and often is not considered in design. The U. S. Army Coastal Engineering Research Center[22] presents an analysis based on several simplifying assumptions. It is assumed that after the wave breaks, it travels forward to its theoretical limit of uprush with an initial velocity of the wave at breaking given by $V = \sqrt{g d_b}$. It is also assumed that the initial velocity and thickness of water mass decrease linearly to zero at the limit of uprush. With these assumptions, the mass and velocity of the water at the point where it intersects the structure can be used to calculate the kinetic energy, which in turn can be converted to an equivalent hydrostatic pressure.

Rubble Mound Structures

Rubble mound type structures, which usually consist of successive layers of graded stone and/or armor units reposing on a prepared foundation bed, are employed chiefly as shore protection or harbor protection in the form of breakwaters, jetties, groins, and so on. These structures are porous, thus transmitting and absorbing some wave energy, and are thus termed flexible structures. The philosophy of design for such flexible structures differs in principle from the design of other structure types. Failure of such structures usually comes about gradually over time because of repeated wave attack and accumulated storm damage. As a cost tradeoff, the design of such structures usually assumes some percent damage is allowed for the design wave condition. As previously mentioned (see Section 3.4), the design wave for this class of structure may be the significant wave or possibly the average of the highest 10%. This selection also depends upon the energy content and duration of the design wave spectrum. Because of the many inherent difficulties of calculating forces on such structures or individual elements, the design of such structures concentrates on the stability of the armor stone against being displaced by the waves. Hence, if some percent damage is allowed, it is assumed that displaced armor may be replaced after the storm wave conditions subside. Empirical formulas have been developed that are solved to yield a weight of armor unit, the size of which depends upon whether it is of stone or concrete – and if it is of concrete, its configuration is also important. Resistance to hydrodynamic forces is also developed by the interlocking of units, size, gradation, and method of placement. It is important to trade off the integrity of greater interlocking and gradation with the consequent loss of porosity, as decreased porosity results in increased wave run-up and reflection. The general design of rubble mound structures and breakwaters is covered in the following selected references: 22, 28, 97, 121, 125, and 126.

Floating Structures

Some of the many applications of floating structures were mentioned at the beginning of this subsection. Some aspects of floating break-waters were mentioned in Section 3.3. Kowalski[127] has summarized a wide variety of floating breakwater types, all of which manifest some degree of effectiveness provided that their major dimension is significantly greater than the incident wavelength. Floating breakwaters are generally "transparent" to the longer-period ocean waves, but are gaining in popularity, primarily because of the increase in pleasure boats and small craft, which are berthed in relatively protected waters where shorter-period waves are a menacing problem. Floating break-waters are relatively insensitive to changing water depths and bottom conditions; they are mobile and likely to be cost-effective for certain applications.

Buoys are another important and special class of floating structures. The theory and design of buoys and buoy systems are thoroughly covered by Berteaux.[115]

Other applications of floating structures include semi-submersibles, floating dry docks, and bargelike pontoon structures used to support various types of plant and equipment and/or residences, floating pontoon bridges, and so forth. These structures, except for semi-submersibles, are usually sheltered from damaging waves, but may have to undergo ocean tow from their construction site to final location. In this case, they must be analyzed for wave bending and torsional moments similarly to ship hulls, but the design wave condition will generally be less severe owing to the shortness of exposure and careful selection of the route and time of year of towing to avoid known areas of frequent storm activity. In general, however, the time and route of tow as well as design wave condition may be specified by the insurance underwriter or by some vessel classification society such as The American Bureau of Shipping, Lloyds Register of Shipping, Det Norske, Veritas, etc. The analysis of forces on and motions of floating bodies is beyond the scope of this text and the reader is referred to the standard works in naval architecture such as reference 11.

The design of fixed structures, such as piers and dolphins, to which floating vessels are moored usually provides for mooring line pulls due to wind and current loads and vessel impact during berthing, and often neglects mooring line pull or breasting forces due to vessel motion in waves. This practice is rationalized in most cases because the berths are usually sheltered from damaging waves, and the vessel would not remain alongside under storm conditions. Wave-induced mooring loads cannot be neglected, however, in exposed or offshore berths. Section 5.4 discusses induced motions in harbors subject to very long-period waves and resonant oscillations. The literature on the effects of waves upon vessels moored alongside piers and wharves is quite sparse. References 28, 128, and 129 provide some practical guidance for estimating wave-induced forces on moored vessels. Wiegel et al.[130] conducted model experiments on a moored model of a converted Liberty ship. The model, which is representative of a typical general cargo type vessel, was moored alongside with fiber lines and fendering devices called camels, as shown in Fig. 3.57, and exposed to head, beam, stern, and quartering seas of varying periods and up to about 5-foot height. Some of the sample record runs for long, intermediate, and short period waves are also shown in Fig. 3.57. Wiegel and his colleagues noted that the higher the wave height, the more regular the force records became; and the shorter the period, the higher was the wave required to cause a uniform force record. The results of other model studies indicate that the force may start to increase exponentially with wave height with higher wave heights. O'Brien and Kuchenreuther[131] reported on some actually measured forces induced in a large vessel by long period waves, and Wilson[132] has analyzed an actual case of critical surging of a moored oil tanker.

More recently Lee et al.[133] have presented a method of predicting the dynamic response of super-tankers moored at a sea berth excited by waves. Lee's findings include the relative effects of fender system

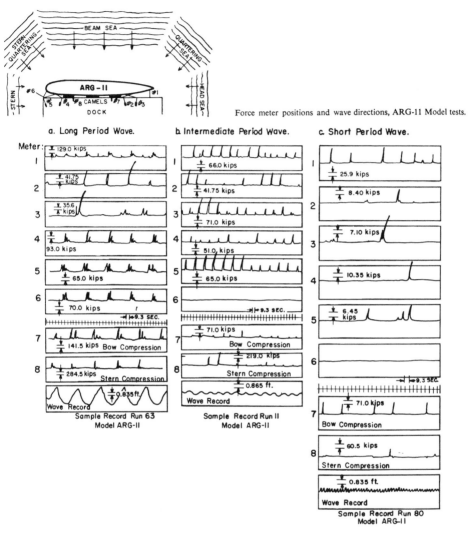

Force meter positions and wave directions, ARG-11 Model tests.

Fig. 3.57. Sample mooring line and camel force records on docked Liberty ship (ARG-11). (From Wiegel, Dilley, and Williams, (1959)[130])

and mooring line characteristics as well as wave characteristics upon the resulting forces and motions.

A structure moored in regular waves will experience a constant horizontal drift force and yaw moment in addition to the oscillatory wave forces. Løken and Olsen[134] have explored the importance of wave drift and slowly varying wave forces in the design of mooring systems.

3.7 OTHER EFFECTS OF WAVES

Wave Run-Up

Wave run-up is the vertical height above the S.W.L. to which water (not spray) will rise when an incident wave impinges upon a structure or shoreline. The run-up height (R) is especially important when determining the heights of seawalls and shore protection structures to preclude overtopping or to reduce overtopping to acceptable limits. The run-up depends upon the structure (or beach) type, slope and surface roughness, the bottom slope seaward of the structure, and the incident wave characteristics, in particular the relative steepness (H/T^2). Figure 3.58 illustrates some of the parameters used in the discussion of run-up. Theory exists for the case of run-up of unbroken waves, but for breaking waves the phenomenon is more complex, and in general the run-up is determined from empirical formulas based upon laboratory investigations. Saville[135] conducted laboratory

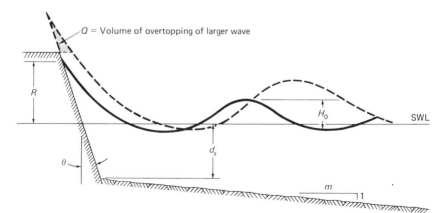

Fig. 3.58. Definition sketch, wave run-up.

experiments and presented the results of his own and of other test data in the form of nondimensional curves in terms of relative run-up, R/H_0, for various structure slopes and wave steepnesses. Such curves, covering a wide range of relative depths and conditions, can be found in the SPM.[22] Instructions for computing run-up on compound slopes and for estimating volumes of overtopping may also be found in the SPM. As a rule of thumb for general preliminary estimates, a value of $R/H_0 = 1.5$ has been used, although R/H_0 for most structures is in the range of 0.9 to 2.0. The relative run-up, however, is quite variable. In general, steeper waves have a lower relative run-up, and steeper structure slopes have a higher relative run-up. The highest relative run-up for steep waves occurs on a slope of approximately 1 on 2, while the highest relative run-up for waves of low steepness occurs on slopes on the order of 1 on 5. It should be borne in mind, however, that most of the available graphs and theories are based upon regular waves. In a train of irregular waves, a long and high wave which is preceded by a shorter and lower wave will have a higher run-up than if preceded by a wave of the same dimensions, as the back draught of the preceding wave affects the run-up of the following wave. Also, the smaller waves tend to be destroyed by the back rush of larger ones so that there are usually fewer run-up crests than incident waves.

The effect of surface roughness on relative run-up was investigated by Battjes,[136] who presented coefficients for various slope facing materials. The relative run-up may be reduced by about 50% for armor stone and rip-rap facing and 10% to 15% for concrete slabs or grass, respectively, from the run-up for a smooth impermeable face for which most experimental data are presented. Savage[137] presented experimental data for rough, permeable slopes. LeMéhaute et al.[138] have summarized the state of the art with regard to wave run-up for both breaking and nonbreaking periodic waves.

Wave Agitation and Reflection in Harbors

Waves propagating into an enclosed and/or semi-enclosed harbor or bay may be reflected at the shoreline and set up complex, confused standing wave patterns within the harbor. They may also initiate or interact with longer period resonant motions within the basin (see Section 5.4). The wakes of vessels transiting the harbor also exacerbate this situation, which is a particularly important consideration in the planning of small craft harbors.[139]   In general, it is desirable to destroy the wave energy as it enters a harbor.   Valembois[140] presented several configurations of harbor "resonators" designed to trap and/or destroy the energy of potentially damaging or noisome waves entering a harbor. Other methods of minimizing this problem during the harbor planning stage include providing gradually sloping beaches where possible to minimize reflection, and avoiding steep, vertical walls with high reflectivity.   LeMehaute[141] has given a thorough treatment of the theory of wave agitation in harbors.

Waves Caused by Vessels

Within harbors and along protected coastlines waves produced by the wakes of large vessels traveling at relatively high speed may be very damaging (e.g., by inducing motion in other moored vessels, causing shoreline erosion, or interfering with recreational or a variety of commercial activities).   In addition, vessels passing close to other moored vessels, even at slow speeds, cause mooring line and breasting forces due to suction and pressure gradients.

The general wave pattern produced by a moving vessel is shown in Fig. 3.59.   Vessel wave patterns are discussed in detail in reference 11 or any basic naval architecture text.   Briefly, a pattern of transverse and diverging waves is produced.   The mechanics of these waves are

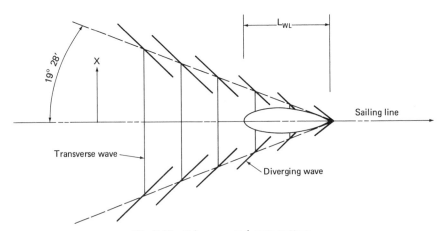

Fig. 3.59. Ship generated wave pattern.

the same in principle as for the wind waves previously discussed in this chapter. As the vessel increases its speed, the length between diverging wave crests increases, to a practical limit, for full-displacement type hulls, to the vessel's water line length. This speed is known as hull speed, and vessels that travel faster than this limit, given by:

$$V_{max} = 1.34 \sqrt{L.W.L.}$$

where V_{max} is in knots, and L.W.L. is the vessel's water line length in feet, are considered to be mathematically "planing" or leaving their bow wave behind, at the expense of a relatively high input of energy. In shallow water, since $C = \sqrt{gd}$, the vessel is virtually forced to travel at a wave speed governed by water depth. This occurs at Froude number, N_f, in excess of approximately 0.6, where:

$$N_f = \frac{V}{\sqrt{gd}}$$

The maximum height of a ship-generated wave occurs at approximately $N_f = 1.0$. The maximum heights of ship-generated waves in harbors can be estimated from Fig. 3.60, after the work of Johnson,[142] which

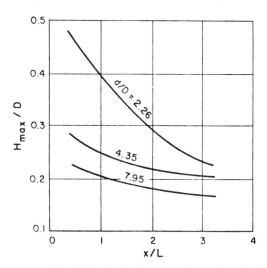

Fig. 3.60. Maximum wave height-ship draft ratio as function of relative distance from sailing line with depth-draft ratio as parameter. (After Johnson[142])

gives the wave-height-to-vessel-draft ratio H_{max}/D versus distance from sailing line to ship length, X/L ratio, with water-depth-to-vessel-draft ratio d/D as a parameter. Note that for the shallower water depths, corresponding to higher N_f, the wave height is larger, but that the rate of decrease in height with distance from the sailing line is also greater. Sorenson[143] found that the rate of decrease in height of the diverging waves is inversely proportional to approximately the cube root of the distance from the sailing line. Table 3.10, after Sorenson,[143] presents a summary of ship wave data for various vessel types. There is a fairly large body of literature on this subject, and the existing knowledge of ship waves is sufficient that for a given vessel, traveling at a known

Table 3.10. Selected Ship-generated wave heights.

Vessel type (1)	Length, in feet (meters) (2)	Beam, in feet (meters) (3)	Draft, in feet (meters) (4)	Displacement, in tons (kilograms) (5)	Water depth, in feet (meters) (6)	Speed, in knots (meters per second) (7)	DISTANCE FROM SAILING LINE, IN FEET (METERS)		
							100 (30.5) H_{max}, in feet (meters) (8)	500 (152.4) H_{max}, in feet (meters) (9)	1,000 (304.8) H_{max}, in feet (meters) (10)
Cabin cruiser[a]	23 (7.0)	8.3 (2.5)	1.7 (0.5)	3 (2,722)	40 (12.2)	6 (3.1)	0.7 (0.2)	0.4 (0.1)	
						10 (5.1)	1.2 (0.4)	0.8 (0.2)	
Coast guard[a] cutter	40 (12.2)	10 (3.0)	3.5 (1.1)	10 (9,072)	38 (11.6)	6 (3.1)	0.6 (0.2)	1.0 (0.3)	
						10 (5.1)	1.5 (0.5)		
						14 (7.2)	2.4[c] (0.7)		
Tugboat[a]	45 (13.7)	13 (4.0)	6 (1.8)	29 (26,309)	37 (11.3)	6 (3.1)	0.6 (0.2)	0.3 (0.1)	
						10 (5.1)	1.5 (0.5)	0.9 (0.3)	
Reconverted air-sea rescue vessel[a]	64 (19.5)	12.8 (3.9)	3 (0.9)	35 (31,752)	40 (12.2)	6 (3.1)	0.3 (0.1)		
						10 (5.1)	1.4 (0.4)	0.8 (0.2)	
						14 (7.2)	2.0[c] (0.6)	1.1[c] (0.3)	
Fireboat[a] (reconverted tug)	100 (30.5)	28 (8.5)	11 (3.4)	343 (311,170)	39 (11.9)	6 (3.1)	0.4 (0.1)	0.2 (0.1)	
						10 (5.1)	1.7 (0.5)	1.0 (0.3)	
						14 (7.2)	3.1 (0.9)	2.6 (0.8)	
Barge[b]	263 (80.2)	55 (16.8)	14 (4.3)	5,420 (4,917,000)	42 (12.8)	10 (5.1)	1.4 (0.4)	0.7 (0.2)	0.3 (0.1)
Moore dry dock tanker[b]	504 (153.6)	66 (20.1)	28 (8.5)	18,800 (17.1×10^6)	56 (17.1)	14 (7.2)		1.5 (0.5)	1.1 (0.3)
						18 (9.3)		5.2 (1.6)	4.7 (1.4)

a Source–Ref.
b Source–Reference
c Vessel beginning to plane.

After Sorenson.[143]

speed in a given water depth, the height of wave at some distance from the sailing line can be readily predicted.

Fatigue of Structures in Waves

Because of the cyclic nature of wave loadings, fatigue stresses must be considered in the design of offshore structures. The magnitude of the load reversals, however, varies greatly in a seaway, unlike the uniform cyclic stresses considered in the design of bridges or buildings, for example. Determination of the number of cycles and their relative magnitude must be based on spectral techniques. For this application, the works of Bretschneider[15] and Korvin-Kroukovsky[12] are especially useful. Riggs[44] demonstrates the application of sea spectra to fatigue considerations for semi-submersible design. More traditional ocean structures, such as piers and other structures, are not usually designed to fatigue criteria per se, but the conservatism of design and allowable stresses perhaps has compensated for the lack of knowledge in this area. The offshore structure design codes,[47, 60] already introduced, cite the importance of fatigue and require welding and detailing of joints and connections to be consistent with sound practice known to prolong fatigue life. Maddox[41] has addressed the problem of fatigue in fixed offshore platforms. At present there does not seem to be a wholly satisfactory criterion for determining allowable stress versus cycles to failure that is meaningful in light of the variability of wave loadings. Figure 3.61 illustrates the relationship of sea spectra and fatigue design considerations.

Scour and Deposition

Wave action may cause scouring of loose bottom material, especially fine sands and silts, which may in turn affect the stability of a structure's foundation. Some criteria for scouring are given in Section 4.4 with regard to current action. Where scouring may be a problem, it should be accounted for in design, by either providing scour protection or assuming that a given amount of bottom material will be removed by scour. Model tests may be useful in this regard. Apelt and MacKnight[85] describe model tests and the final method of scour protection provided for a large offshore caisson-type berthing structure off the Australian coast.

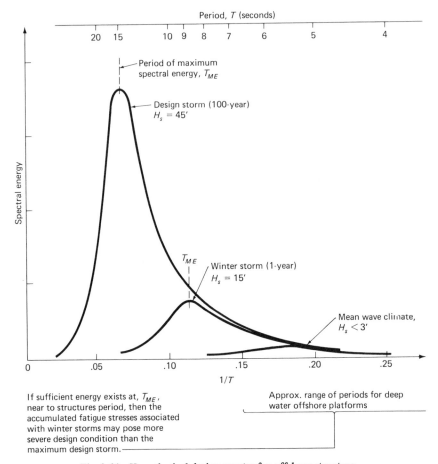

Fig. 3.61. Hypothetical design spectra for offshore structure.

REFERENCES

1. Airy, G. B. "On Tides and Waves," *Encyclopedia Metropolitania*, London, 1845.
2. Gaythwaite, J. W. "Structural Design Considerations in the Marine Environment," *Boston Soc. of Civil Eng'rs. Section, ASCE*, Vol. 65, No. 3, Oct. 1978.
3. Phillips, O. M. "On the Generation of Waves by Turbulent Wind," *Journal of Fluid Mech's.*, 2, Part 5, 1957.
4. Myers, J. J., Holm, C. H., and McAllister, ed. *Handbook of Ocean and Underwater Engineering*, McGraw-Hill, New York, 1969.

5. Bretschneider, C. L. "Revised Wave Forecasting Relationships," *Proc.* 2nd. Conf. on Coastal Eng'g., Council on Wave Research, 1952.
6. U. S. Naval Oceanographic Office. *Techniques for Forecasting Wind, Waves and Swell*, H.O. No. 604, Washington, D.C., 1951.
7. Saville, T., Jr. *The Effect of Fetch Width on Wave Generation*, TM-70, U. S. Army, Beach Erosion Board, 1954.
8. Bretschneider, C. L. "Hurricane Design Wave Practices," *Proc ASCE*, WW-2, Vol. 83, May 1957.
9. Longuet-Higgins, M. S. "On the Statistical Distribution of the Heights of Sea Waves," *Journal of Marine Science*, Vol. 2, 1952.
10. St. Denis, M., and Pierson, W. J., "On the Motions of Ships in Confused Seas," *Trans. SNAME*, Vol. 61, 1953.
11. Comstock, J. P., ed. *Principles of Naval Architecture*, SNAME, New York, 1967.
12. Korvin-Kroukovsky, B. V. *Theory of Seakeeping*, SNAME, 1961.
13. Michel, W. H. "Sea Spectra Simplified," *Marine Technology Journal*, SNAME, New York, Jan. 1968.
14. Pierson, W. J., and Moskowitz, L. *A Proposed Spectral Form for Fully Developed Wind Seas Based on the Similarity Theory of S. A. Kitaigorodskii*, USNOO, T.R. cont. No. N62306-1042, 1963.
15. Bretschneider, C. L. *Wave Variability and Wave Spectra for Wind-Generated Gravity Waves*, U. S. Army, Beach Erosion Board, TM-118, 1959.
16. Bretschneider, C. L., and Tamaye, E. E. "Hurricane Wind and Wave Forecasting Techniques," *Proc ASCE*, 15th Conf. on Coastal Eng'g., 1976.
17. Wilson, B. W. *Graphical Approach to the Forecasting of Waves in Moving Fetches*, USACE, Beach Erosion Board, TM-73, 1955.
18. Sverdrup, H. U., and Munk, W. H. *Wind, Sea, and Swell: Theory of Relations for Forecasting*, U. S. Navy Hydrographic Office, H.O. 601, 1947.
19. Bretschneider, C. L. "Revisions in Wave Forecasting: Deep and Shallow Water," *Proc. ASCE*, 6th Conf. on Coastal Eng'g, Council on Wave Research, 1958.
20. Pierson, W. J., Neumann, G., and James, R. W. *Practical Methods for Observing and Forecasting Ocean Waves by Means of Wave Spectra and Statistics*, USNOO, H.O. 603, 1955.
21. Neumann, G. *An Ocean Wave Spectra and a New Method of Forecasting Wind Generated Sea*, USACE, Beach Erosion Board, TM-43, 1952.
22. U. S. Army. *Shore Protection Manual*, 3 vols., USACE, Coastal Engineering Research Center, 1977.
23. Kinsman, B. *Wind Waves*, Prentice-Hall, Englewood Cliffs, N.J., 1965.
24. Bretschneider, C. L., "Wave Generation by Wind, Deep and Shallow Water," in *Estuary and Coastline Hydrodynamics*, edited by A. T. Ippen, McGraw-Hill, New York, 1966.
25. Wiegel, R. L. *Gravity Waves: Tables of Functions*, University of California, Council on Wave Research, Berkeley, 1954.
26. Bretschneider, C. L. and Reid, R. O. *Changes in Wave Height Due to Bottom Friction, Percolation, and Refraction*, USACE, BEB, TM-45, 1954.

27. Johnson, J. W., O'Brien, M. P., and Isaacs, J. D. *Graphical Construction of Wave Refraction Diagrams*, USNOO, H.O. 605, 1948.

28. Wiegel, R. L. *Oceanographical Engineering*, Prentice-Hall, Englewood Cliffs, N.J., 1964.

29. Sorenson, R. M. *Basic Coastal Engineering*, John Wiley & Sons, New York, 1978.

30. Wiegel, R. L. "Diffraction of Waves by Semi-infinite Breakwater," *Journal Hydraulics Div., ASCE*, Vol. 88, 1962.

31. Penny, W. G., and Price, A. T. "The Diffraction Theory of Sea Waves by Breakwaters and the Shelter Afforded by Breakwaters," *Philosophical Trans. Royal Soc., Series A, London*, 1952.

32. Johnson, J. W. "Generalized Wave Diffraction Diagrams," *Proc. ASCE*, 2nd Conf. Coastal Eng'g., Council on Wave Research, 1952.

33. Ibid. "Engineering Aspects of Diffraction & Refraction," *Trans. ASCE*, Vol. 118, 1953.

34. Dean, R. G., and Harleman, D. R. F. "Interaction of Structures and Waves," in *Estuary and Coastline Hydrodynamics*, edited by A. C. Ippen, McGraw-Hill, New York, 1966.

35. Kincaid, G. A. "Effects of Natural Period Upon the Characteristics of a Moored Floating Breakwater," M.I.T., B. Sci. thesis, Dept. of C. E., 1960.

36. Dalrymple, R. A., and Dean, R. G. "Waves of Maximum Height on Uniform Currents," *Proc. ASCE*, WW-3, Vol. 101, Aug. 1975.

37. Herbich, J. B., and Hales, L. "The Effect of Tidal Inlet Currents on the Characteristics and Energy Propagation of Ocean Waves," *OTC* paper No. 1618, Houston, 1972.

38. Johnson, J. W. "The Refraction of Surface Waves by Currents," *Trans. Am. Geophysical Union*, Vol. 28, Dec. 1947.

39. Wiegel, R. L., and Beebe, K. E. "The Design Wave in Shallow Water," *Proc. ASCE*, WW-1, Vol. 82, March 1956.

40. U. S. Navy. *Breakers & Surf: Principles in Forecasting*, U. S. Navy Hydrographic Office, H.O. 234, 1944.

41. Maddox, N. R. "Fatigue Analysis for Deepwater Fixed Bottom Platforms," *OTC* #2051, May 1974.

42. Nolan, W. C., and Honsinger, V. C. "Wave Induced Vibrations of Offshore Structures," *Trans. SNAME*, New York, 1962.

43. Wirsching, P. H., and Prasthofer, P. H. "Preliminary Dynamic Assessment of Deepwater Platforms," *Proc. ASCE*, ST-7, Vol. 102, July 1976.

44. Riggs, R. P. "Fatigue Considerations for Semi-submersible Structures," *Marine Technology*, SNAME, Vol. 16, No. 1, Jan. 1979.

45. Weibull, W. "A Statistical Distribution Function of Wide Applicability," *ASME*, Applied Mech., Div., Ann. Mtg., 1951.

46. Bretschneider, C. L. "Wind and Wave Loads," in *Handbook of Ocean and Underwater Engineering*, edited by Myers, J. J., McAllister, R. F., and Holm, C. H., McGraw-Hill, New York, 1969.

47. Det Norske Veritas. "Rules for the Design, Construction and Inspection of Fixed Offshore Structures," Oslo, 1974.
48. Ochi, M. K. "Wave Statistics for the Design of Ships and Ocean Structures," *Trans. SNAME*, Vol. 86, 1978.
49. Collins, J. I. "Wave Statistics from Hurricane Dora," *Proc. ASCE*, WW-2, Vol. 93, May 1967.
50. Chakrabarti, S. K., and Cooley, R. P. "Ocean Wave Statistics for 1961 North Atlantic Storm," *Proc. ASCE*, WW-4, Vol. 103, Nov. 1977.
51. Sellars, F. "Maximum Wave Statistics for Design," *T&R Bulletin* No. 1-37, SNAME, March 1978.
52. "How Big Can Waves Get?," anonymous article in *Ocean Industry Magazine* (pp. 40–41), Houston, June 1978.
53. Hogben, N., and Lumb, F. E. *Ocean Wave Statistics*, National Physical Laboratory, H. M. Stationery Off., London, 1967.
54. Roll, H. U. "Height, Length and Steepness of Sea Waves in the North Atlantic and Dimensions of Sea Waves as Functions of Wind Force," *T&R Bulletin* No. 1-19, SNAME, 1953.
55. Yamanouchi, Y., and Ogawa, A. *Statistical Diagrams on the Winds and Waves on the North Pacific Ocean*, Ship Research Institute of Japan, 1970.
56. Ward, E. G., Evans, D. J., and Pompa, T. A. "Extreme Wave Heights along the Atlantic Coast of the United States," *OTC* No. 2846, Houston, May 1977.
57. Bea, R. G. "Gulf of Mexico Hurricane Wave Heights," *OTC* No. 2110, Houston, May 1974.
58. Resio, D. T., and Hiipakka, L. W. "Great Lakes Wave Information," *Proc. ASCE*, 15th Conf. on Coastal Eng'g, Vol. 0, 1976.
59. Thompson, E. F., and Harris, D. L. "A Wave Climatology for U. S. Coastal Waters," *OTC* No. 1693, Houston, May 1972.
60. American Petroleum Institute, *Recommended Practice for Planning, Designing, and Constructing Fixed Offshore Platforms*, API, RP-2A, 10th ed., Dallas, March 1979.
61. Galvin, C. J. "Breaker Travel and Choice of Design Wave Height," *Proc. ASCE*, WW-2, Vol. 95, May 1969.
62. Weggel, J. R. "Maximum Breaker Height," *Proc. ASCE*, WW-4, Vol. 98, Nov. 1972.
63. Bretschneider, C. L. "Selection of Design Wave for Offshore Structures," *Proc. ASCE*, WW-2, Vol. 84, March 1958.
64. Lamb, Sir Horace. *Hydrodynamics*, Cambridge University Press, 1932.
65. Eagleson, P. S., and Dean, R. G. "Small Amplitude Wave Theory" and "Finite Amplitude Waves," in *Estuary and Coastline Hydrodynamics*, edited by A. T. Ippen, McGraw-Hill, New York 1966.
66. McCormick, M. E. *Ocean Engineering Wave Mechanics*, John Wiley & Sons, New York, 1973.
67. LeMéhaute, B. *An Introduction to Hydrodynamics and Water Waves*, Water Wave Theories, Vol. II, TR ERL 118-POL-3-2, U. S. Dept. Comm., ESSA, Washington, D.C., 1969.

68. Ursell, F. "Mass Transport in Gravity Waves," *Proc. Cambridge Philosophical Society*, Vol. 49, Jan. 1953.
69. Stokes, G. G. *On the Theory of Oscillatory Waves*, Math. and Physical Papers, Vol. I, Cambridge University Press, 1880.
70. Skjelbreia, L., and Hendrickson, J. A. *Fifth Order Gravity Wave Theory and Tables of Functions*, National Engineering Science Company, Washington, D.C. 1962.
71. Michell, J. H. "On the Highest Waves in Water," *Philosophical Magazine*, 5th series, Vol. 36, 1893.
72. Havelock, T. H. "Periodic Irrotational Waves of Finite Height," *Proc. Royal Soc. London*, Vol. 95, No. A665, 1918.
73. Dean, R. G. "Stream Function Representation of Non-Linear Ocean Waves," *Journal of Geophysical Research*, Vol. 70, No. 18, Sept. 1965.
74. Ibid. "Relative Validities of Water Wave Theories," *Proc. ASCE*, Civil Eng'g. in the Oceans I, Sept. 1967.
75. Monkmeyer, P. L. "Higher Order Theory for Symmetrical Gravity Waves," *Proc. ASCE*, 12th Conf. on Coastal Eng'g., 1970.
76. Dean, R. G. "Evaluation and Development of Water Wave Theories for Engineering Application," 2 vols. USACE, CERC, S.R. No. 1, Nov. 1974.
77. Masch, F. D., and Wiegel, R. L. *Cnoidal Waves, Tables of Functions*, Council on Wave Research, The Eng'g Fdn., Richmond, Calif., 1961.
78. Keulagan, G. H., and Patterson, G. W. *Mathematical Theory of Irrotational Translation Waves*, R.P. No. 1272, National Bureau of Standards, Washington, D.C., 1940.
79. Munk, W. H. "The Solitary Wave Theory and Its Application to Surf Problems," *Annals of the N.Y. Academy of Sciences*, Vol. 51, 1949.
80. Daily. J. W., and Stephan, S. C. "Characteristics of a Solitary Wave," *Trans. ASCE*, Vol. 118, 1953.
81. Iwasa, Y. "Analytical Consideration on Cnoidal and Solitary Waves," *Memoirs* of the Faculty of Eng'g., Kyoto University, Japan, 1955.
82. Gerstner, F. J. "Theorie die Wellen," *Ann. Physik*, 32, 1809.
83. American Concrete Institute. "Guide for the Design and Construction of Fixed Offshore Concrete Structures," Report of ACI Committee 357, No. 75–72, Dec. 1978.
84. Plate, E. J., and Nath, J. H. "Modeling of Structures Subjected to Wind Waves," *Proc. ASCE*, WW-4, Vol. 95, Nov. 1969.
85. Apelt, C. J., and MacKnight, A. "Wave Action on Large Offshore Structures," *Proc. ASCE*, 15th Conf. on Coastal Eng'g., Honolulu, 1976.
86. Wiegel, R. L., Beebe, K. E., and Moon, J. "Ocean Wave Forces on Circular Cylindrical Piles," *Proc. ASCE, Journal of Hydraulics Div.*, April 1957.
87. Reid, R. O. "Correlation of Water Level Variations with Wave Forces on a Vertical Pile for Non-periodic Waves," *Proc.* 6th Conf. on Coastal Eng'g., Council on Wave Research, 1958.
88. Wilson, B. W. *Analysis of Wave Forces on a 30 inch Diameter Pile Under Confused Sea Conditions*, U. S. Army, CERC, T.M. #15, Washington, D.C., 1965.

89. Evans, D. J. "Analysis of Wave Force Data," *Proc. OTC*, paper No. 1005, Houston, 1969.

90. A Agard, P. M., and Dean, R. G. "Wave Forces: Data Analysis and Engineering Calculation Method," *Proc. OTC*, paper No. 1008, Houston, 1969.

91. Morison, J. R., Johnson, J. W., O'Brien, M. P., and Schaaf, S. A. "The Forces Exerted by Surface Waves on Piles," *Petroleum Transactions*, American Inst. of Mining Eng'rs., Vol. 189, 1950.

92. Reid, R. O., and Bretschneider, C. L. "The Design Wave in Deep or Shallow Water, Storm Tide, and Forces on Vertical Piling and Large Submerged Objects," Dept. of Oceanography, Texas A & M Univ., College Station, Texas, 1953.

93. Skjelbria, L., et al. *Loading on Cylindrical Piling Due to the Action of Ocean Waves*, 4 vols., U. S. Navy, Civil Eng'g. Lab., 1960.

94. Dean, R. G. "Stream Function Representation of Non-Linear Ocean Waves," *Journal of Geophysical Research*, Vol. 70, No. 18, 1965.

95. Goda, Y. *Wave Force Experiments on a Vertical Circular Cylinder: Experiments and a Proposed Method of Wave Force Computation*, Ministry of Japan, Port, Harbor, Tech. Res. Inst., T. R. 8, Yokosuka, 1964.

96. Morison, J. R., Johnson, J. W., and O'Brien, M. P. "Experimental Studies of Forces on Piles," *Proc.* 4th Conf. on Coastal Eng'g., Council on Wave Research, 1953.

97. Quinn, A. D. F. *Design and Construction of Ports and Marine Structures*, McGraw-Hill, New York, 1972.

98. Hall, M. A. *Laboratory Study of Breaking Wave Forces on Piles*, U. S. Army Corps of Engineers, BEB, TM-106, 1958.

99. Ross, C. W. *Large Scale Tests of Wave Forces on Piling*, USACE, BEB, TM-111, 1959.

100. Keulegan, G. H., and Carpenter, L. H. *Forces on Cylinders and Plates in an Oscillating Fluid*, National Bureau of Standards, NBS Report #4821, 1956.

101. Bidde, D. D. *Wave Forces on a Circular Pile Due to Eddy Shedding*, University of California, HEL 9-16, Berkeley, 1970.

102. Nath, J. H., and Harleman, D. R. F. "Dynamics of Fixed Towers in Deep-Water Random Waves," *Proc. ASCE*, WW-4, Vol. 95, Nov. 1969.

103. Brannon, H. R., Loftin, T. D., and Whitfield, J. H. "Deepwater Platform Design," *OTC* paper No. 2120, Houston, May 1974.

104. Malhotra, A. K., and Penzien, J. "Response of Offshore Structures to Random Wave Forces," *Proc. ASCE*, ST-10, Vol. 96, Oct. 1970.

105. MacCamy, R. C., and Fuchs, R. A. *Wave Forces on Piles: A Diffraction Theory*, USACE, BEB, TM-69, 1954.

106. Mogridge, G. R., and Jamieson, W. W. *A Design Method for the Estimation of Wave Loads on Square Caissons*, National Research Council of Canada, Hydraulics Laboratory, T.R. #LTR-HY-57, 1976.

107. Milgram, J. H., and Halkyard, J. E. "Wave Forces on Large Objects in the Sea," *Journal of Ship Research*, SNAME, Vol. 15, No. 2, June, 1971.

108 Garrison, C. J., and Chow, P. Y. "Wave Forces on Submerged Bodies," *Proc. ASCE*, WW-3, Vol. 98, Aug., 1972.

109. Hogben, N., and Standing, R. G. "Wave Loads on Large Bodies," *Int. Symposium on the Dynamics of Marine Vehicles and Structures in Waves*, Mech. Eng'g. Publications, Ltd., London, 1975.
110. Chakrabarti, S. K. "Wave Forces on Submerged Objects of Symmetry," *Proc. ASCE*, WW-2, Vol. 99, May 1973.
111. Vongvisessomjai, S., and Silvester, R. "Wave Forces on Submerged Objects," *Proc. ASCE*, 15th Conf. on Coastal Eng'g., Honolulu, 1976.
112. Brown, R. J. "Hydrodynamic Forces on a Submarine Pipeline," *Proc. ASCE, Journal of Pipeline Div.*, March 1967.
113. Helfinstine, R. A., and Shupe, J. W. "Lift and Drag on a Model Offshore Pipeline," *OTC* paper #1568, Houston, May 1972.
114. Braten, E. F., and Wallace, R. "Wave Forces on Submerged Pipelines," *Proc. ASCE*, 13th Conf. on Coastal Eng'g., Vancouver, B.C., 1972.
115. Berteaux, H. O. *Buoy Engineering*, John Wiley & Sons, New York, 1976.
116. Wilson, B. W., and Garbaccio, D. H. "Dynamics of Ship Anchor Lines in Waves and Currents," *Proc. ASCE*, Civil Eng'g. in the Oceans, San Francisco, 1967.
117. Chang, P. Y., and Pilkey, W. D. "The Analysis of Mooring Lines," *OTC* paper #1502, Houston, May 1971.
118. Hudson, R. Y. "Wave Forces on Breakwaters," *Trans. ASCE*, Vol. 118, 1953.
119. Sainflou, M. "Treatise on Vertical Breakwaters," *Annals des Ponts et Chaussees*, Paris, 1928.
120. Molitor, D. A. "Wave Pressures on Seawalls and Breakwaters," *Trans. ASCE*, Vol. 100, 1935.
121. Minikin, R. R. *Winds, Waves and Maritime Structures*, Charles Griffin & Co., Ltd., London, 1963.
122. Nagai, S. "Wave Forces on Vertical Wall Breakwaters in Deepwater," appendix 6, *Port Engineering*, by Per Bruun, Gulf Publishing Co., Houston, 1976.
123. Bagnold, R. A. "Interim Report on Wave Pressure Research," *Journal of the Insititute of Civil Eng'rs.*, Vol. 12, London, 1939.
124. Denny, D. F. "Further Experiments on Wave Pressures," *Journal of the Institute of Civil Eng'rs.*, Vol. 35, London, 1951.
125. Bruun, P. *Port Engineering*, Gulf Publishing Co., Houston, 1976.
126. Silvester, R. *Coastal Engineering*, 2 vols., Elsevier Scientific Pub. Co., 1974.
127. Kowalski, T. *Floating Breakwaters Conference Papers*, University of Rhode Island, M.T.R. #24, Kingston, R.I., 1974.
128. U. S. Navy. *Harbor and Coastal Facilities*, DM-26, Naval Fac. Eng'g. Comm., July 1968.
129. *Analytical Treatment of Problems of Berthing and Mooring Ships*, NATO, Advanced Study Inst., ASCE, New York, 1970.

130. Wiegel, R. L., Dilley, R. A., and Williams, J. B. "Model Study of Mooring Forces of Docked Ship," *Proc. ASCE*, WW-2, Vol. 85, June 1959.

131. O'Brien, J. T., and Kuchenreuther, D. I. "Forces Induced on a Large Vessel by Surge," *Proc. ASCE*, WW-2, Vol. 84, March 1958.

132. Wilson, B. W. "A Case of Critical Surging of a Moored Ship," *Proc. ASCE*, WW-4, Vol. 85, Dec. 1959.

133. Lee, T. T., Nagai, S., and Oda, K. "On the Determination of Impact Forces, Mooring Forces & Motions of Super-Tankers at Marine Terminal," *OTC* paper #2211, Houston, May 1975.

134. Loken, A. E., and Olsen, O. A. "The Influence of Slowly Varying Wave Forces on Mooring Systems," *OTC* paper #3626, Houston, May 1979.

135. Saville, T. "Wave Run-Up on Composite Slopes," *Proc.* 6th Conf. on Coastal Eng'g., Council on Wave Research, Berkeley, 1957.

136. Battjes, J. A. Discussion of, "The Run-Up of Waves on Sloping Faces – A Review of the Present State of Knowledge," by N. B. Webber and G. N. Bullock, *Proc. Conf. on Dynamics of Waves in Civil Eng'g.*, Wiley, New York, 1970.

137. Savage, R. P. "Wave Run-Up on Roughened and Permeable Slopes," *Proc. ASCE*, WW-3, Vol. 84, May 1958.

138. LeMehaute, B., Koh, R. C. Y., and Li-San Hwang. "A Synthesis on Wave Run-Up," *Proc. ASCE*, WW-1, Vol. 94, Feb. 1968.

139. Ibid. *Report on Small Craft Harbors*, M & R #50, ASCE, New York, 1969.

140. Valembois, J. "Investigation of the Effect of Resonant Structures on Wave Propagation," *Proc.* Minn, Int. Hyd. Convention, Minneapolis, Sept. 1953.

141. LeMehaute, B. "Theory of Wave Agitation in a Harbor," *Journal, Hydraulics Div. ASCE*, March 1961.

142. Johnson, J. W. "Ship Waves in Navigation Channels," *Proc.* 6th Conf. on Coastal Eng'g., Council on Wave Research, Berkeley, 1958.

143. Sorenson, R. M. "Water Waves Produced by Ships," *Proc. ASCE*, WW-2, Vol. 99, 1973.

4
Effects of Currents

Currents may be sources of significant loads on marine structures, especially upon moored vessels and when additive to wave loads on offshore structures. Moreover, they generally present other adverse functional effects, such as scouring, deposition, increased corrosion rates, and modification of wave effects; and they cause impact of ice and flotsam and hamper construction operations – to name a few of their effects.

Currents are essentially a horizontal movement of water, which can be due to various factors. There are large-scale ocean surface currents caused by the major wind systems, and there are deep-running sea bottom currents associated with density differences. There are more localized tidal currents, wind stress currents, hydraulic currents, and turbidity currents generated by underwater landslides. Of the above, tidal currents and storm-related wind stress currents are of the greatest concern in structural design.

The various types of currents, their general behavior, modifying effects, and nomenclature are discussed first. Then properties such as extreme velocities and vertical velocity distribution are considered in terms of design criteria. The evaluation of current forces follows the selection of design current data. Finally, effects other than structural loadings are evaluated.

4.1 TYPES OF BEHAVIOR OF OCEAN CURRENTS

General Discussion

Currents can be generally classified by their mode of origin, as tidal currents, wind stress currents, hydraulic currents, turbidity currents, littoral currents, and density currents. All currents are modified by the earth's rotation (coriolis force), friction, and boundary conditions of land and other water masses. Currents range in velocity from a few tenths of a knot for major ocean currents to several knots for localized tidal currents. Current measurements may be based upon eulerian (fixed position) concepts, which observe the flow at a given point as the water particles pass, such as would be recorded by an anchored current meter; or measurements may be based on Lagrangian techniques, which follow the path and behavior of a given particle such as the trajectory followed by a drogue or dye. The methods complement each other, and one or the other type may be more informative for a given application.

Because of the complexities of figuring current behavior, accurate field measurements taken over a sufficient period of time to note important variations are essential in evaluating design parameters at a given site, but field measurement techniques are beyond the scope of this text. The structural engineer can get a general estimation of current velocities for many locations in the world, where significant local currents exist, from the tidal current tables[1] for a given locality, published annually by the National Oceanic and Atmospheric Administration (NOAA). NOAA also publishes tidal current charts for selected locations. Depending upon the nature and importance of the structure and the strength and behavior of currents at the site, the designer may request that measurements be taken to record daily and/or seasonal maximums, the vertical and possibly the horizontal velocity distribution, and the general set and drift of the current as affects sediment transport, scouring and deposition, and other factors to be discussed.

The "set" of a current is the general compass direction toward which it flows, whereas the "drift" is the general surface velocity in knots or in miles per day for slower large-scale currents. Tidal currents "flood" with a rising tide and "ebb" with a falling tide. Races exist where land constricts the flow of water horizontally, and overfalls exist where the bottom rises abruptly and constricts the flow vertically.

The major ocean surface currents are driven by the predominating world wind systems and hence can be considered large-scale interactions with atmospheric movements. These currents form clockwise-moving gyres in the Northern Hemisphere and run counterclockwise in the Southern Hemisphere, owing to deflection by coriolis force.

The principal surface currents also may form branches and eddies.

Coriolis force is a fictitious acceleration caused by the rotation of the earth; it is nil at the equator and has its maximum effect at the poles. The coriolis force can be considered to have a vertical and a horizontal component, but the vertical component is small compared to gravity and is usually neglected. Considering the wind to be the driving force, the horizontal component of the coriolis acceleration is given by:

$$F_{CH} = 2 \, \Omega \, V_w \sin \phi \qquad (4.1)$$

where

Ω = the angular rotation of the earth
    = 7.3×10^{-5} radians/sec
V_w = wind velocity in ft/sec for F in ft/sec²
ϕ = latitude in degrees

The coriolis force always acts at right angles to the wind direction. Wind-driven currents diminish rapidly in velocity with depth and also change direction with depth, as described by the Ekman spiral.

Ekman[2] studied the effects of the earth's rotation on ocean currents assuming only coriolis and friction forces were effective and that eddy viscosity effects were constant with depth. He predicted that the direction of the current would change with depth until some depth was reached, called the frictional layer or Ekman's layer, where the current direction would be exactly opposite to the direction at the surface. Further he predicted that the surface current would act at 45° from the wind direction (to the right in the Northern Hemisphere) and that the net transport of water would be 90° from the wind direction. The ends of the velocity vectors form a logarithmic spiral when projected on a horizontal plane, as illustrated in Fig. 4.1. In reality, the surface current direction is always at some angle less than 45° from the wind and the net transport of water is at some angle less than 90° from the wind direction because of the action of other effects not considered by Ekman, however, his theory forms an important illustrative model of actual current behavior. Rossby and Montgomery[3] developed a boundary layer

theory for wind-induced currents that modified Ekman's work
The structural designer need not delve too deeply into the theory of
ocean currents, but should be aware of this particular aspect of current behavior as it effects more short-term wind-generated currents
that may have significant surface velocities during intense storms.
Such wind stress currents will be discussed later in this chapter.

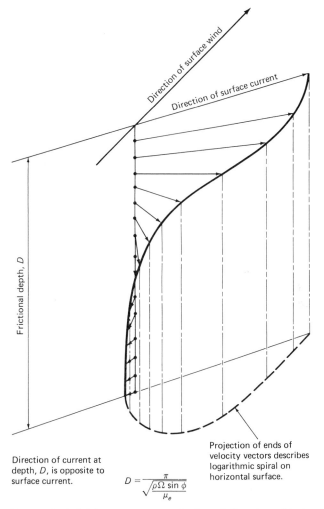

Direction of surface wind

Direction of surface current

Frictional depth, D

Direction of current at
depth, D, is opposite to
surface current.

$$D = \frac{\pi}{\sqrt{\frac{\rho \Omega \sin \phi}{\mu_e}}}$$

Projection of ends of
velocity vectors describes
logarithmic spiral on
horizontal surface.

**Fig. 4.1.   Change of wind-driven current magnitude and direction with depth, Ekman's
spiral.**

The deep-running primary ocean currents, however, are driven by density differences caused by differences in temperature and/or salinity (the thermo-haline circulation). For a given depth, assuming that salinity is constant, the temperature determines the density of the water. As water of different temperatures tends to stay separated in layers with little mixing (thermoclines), a flow is set up such that water of lower density (high water) flows toward water of higher density (low water). Isopycnic lines could be plotted where the maximum differences in higher and lower water might be 1 or 2 feet in 40 miles. Density currents are also modified by coriolis force, and in general the steeper the density gradient the faster the current.

The major ocean currents are of primary interest to oceanographers and meteorologists as they represent a massive transport of water and exert a major influence on world climates. Such currents are not of particular interest here, and are beyond the scope of this book. Interested readers are referred to the general oceanography references given at the end of Chapter 1. The U.S. Naval Oceanographic Office publishes various charts and information on ocean surface currents for different parts of the world.

Turbidity currents are caused by mixing of bottom sediment with seawater, resulting in an increased density of water locally such as may be caused by an underwater landslide. Such down-slope currents are among the fastest known ocean currents, and may occur suddenly and catastrophically. Some turbidity currents have been reckoned to attain local velocities of perhaps 50 knots, although such high values are not confirmed. In 1929 an earthquake occurred near the Laurentian Channel on the Grand Banks of Newfoundland. Underwater cables were broken within 60 miles south of the epicenter, and later a series of cable breaks occurred immediately after a slump, presumably because of the associated large-scale turbidity currents (as reported by Shepard[4]). These currents may be of concern in the siting of certain deep water fixed structures and of underwater pipelines and cables, but there are few empirical data extant, and the problem of predicting their occurrence and severity makes them of more academic than practical interest to the structural designer.

In narrow straits joining water masses of different densities, differential currents and a resulting current shear may exist. In the straits of Gilbralter, for example, water enters the Mediterranean Sea from the Atlantic Ocean at a velocity of greater than 1 knot on the surface, while at a depth of about 300 feet the more saline Mediterranean water is flowing out into the Atlantic at a speed of about 1 knot. Thus a 2-knot current shear is set up at the interface. The reversal of current direction takes place within a relatively thin layer of turbulence. Such a current shear can develop internal waves at the interface of the two waters of different density in much the same way as surface waves are set up between the sea and atmosphere. Current shear can be particularly troublesome to cable and mooring systems, especially if the shear is unexpected.

Littoral currents and their associated longshore and rip currents are caused by waves breaking at an angle to a beach. As sea waves approach shoaling water, they become waves of translation and pile up water against the coast. A hydraulic grade line is thus established, which means that somewhere along the beach there must be a return flow to the sea. Referring to Fig. 4.2, longshore currents are set up inside the breaker zone parallel to the coast, which feed the return flow to the sea, known as rip currents. Littoral currents and processes are of great importance in shoreline protection and certain coastal engineering problems, but are beyond the scope of this text. The U.S. Army Shore Protection Manual[5] provides a thorough treatment of littoral transport problems.

Tidal and wind drift currents are of the most relevance to the structural engineer in the planning and design of fixed and floating structures; so they will be discussed in some detail in this chapter.

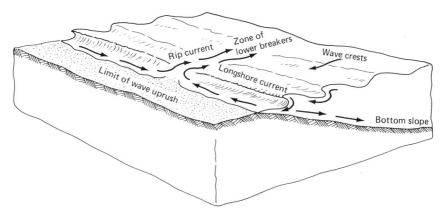

Fig. 4.2. Littoral currents.

Hydraulic currents, which, like river currents, flow because of a difference in water elevations, will be considered as a special case of tidal current here. Because at a given location the actual current observed may be a combination of many of the different types discussed, reliance on actual measurements and published data is necessary. The theoretical prediction of currents and their variations is another area of study that is best left to specialist oceanographers and hydrodynamicists.

Locally, currents may be modified by river run-off and the relative density differences between fresh and salt water. Currents modify waves and themselves are modified by waves (see Section 3.3). Superposition of various currents is an important consideration in selecting design velocities, as will be elaborated upon in this chapter.

Tidal Currents

Tidal currents are a horizontal flow of water associated with the vertical rise and fall of the tides. Offshore, and near unbounded coastlines, tidal currents are rotary in nature, changing speed and direction through all points of the compass during a tidal cycle. In the Northern Hemisphere, these currents change direction in a clockwise manner primarily due to the effects of the earth's rotation, but the rotary nature of the currents may be further complicated by interference due to the differing times and directions of arrival of tidal waves (see Section 5.1).

Figure 4.3 shows a typical current rose, or current ellipse, where the end points of the velocity vectors about a given point approxi-

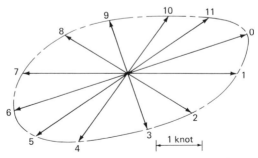

Fig. 4.3. Rotary current (Semi-diurnal tide). Numbers represent hours after high water at reference station.

mate an ellipse. This closed ellipse is typical of rotary currents exhibited by semi-diurnal tides where subsequent highs and lows are about equal in height. Where diurnal inequalities exist (see Section 5.1), the current ellipse may actually resemble an ellipse circumscribed by an ellipse. The effect of bottom friction is to slow down the lower layers of water. There exists a layer of frictional influence near the bottom with a smaller current ellipse and different directions of highest velocity compared to the surface currents. The vertical distribution of current velocities will be discussed further in Section 4.2.

As there are frequently other water movements interacting with tidal currents, the current ellipse can be further modified as shown in Fig. 4.4 to account for local currents due, for example, to strong winds, river run-off, ocean circulation, and so on. In most localities subject to semi-diurnal type tides there is a definite relationship

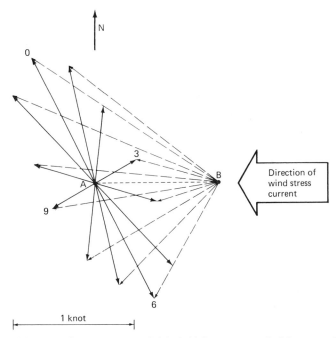

Fig. 4.4. Modification of rotary current. Original tidal current rose (solid arrows with center at "A". For effect of 1-knot wind stress current from the east, distance "A-B" is scaled toward current. All vectors can now be scaled from "B" (dashed arrows) for total combined current.

between the time of high water and the time of maximum currents, so that nautical charts displaying current roses usually reference them to time of high water at some locality. In areas subject to diurnal inequalities, correlation between high water and maximum current is nebulous and should be considered as generally unpredictable. There is not generally any definite correlation between the strength of the current and the height of the tide from locality to locality, but at a given locality currents will be stronger when tides are higher than normal (because of spring tides or storm conditions, for example) in approximate proportion to the difference in tide height. During spring tides (see Section 5.1) current velocities may be on the order of 10% to 20% higher than average, and during neap tides they may be 10% to 20% lower.

In narrow channels or sounds subject to tidal variations the currents are restricted to flow essentially along one path in a reversing manner known as the flood (incoming) and the ebb (outgoing) current. Such a reversing current is depicted in Fig. 4.5 for a semidiurnal tide where the variation in velocity with time approximates a sine wave. Because the flow of water is restricted, reversing currents are typically stronger than rotary currents. Offshore tidal currents may attain a maximum velocity of 1 or possibly 2 knots where reversing currents are typically greater than 1 knot at strength, and may reach velocities of greater than 10 knots under some conditions. Another notable difference between rotary and reversing currents

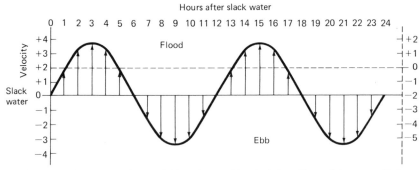

Fig. 4.5. Reversing current for semi-diurnal type tide. Relative effect of river run-off shown by dashed scale (for 2-kt river flow).

is the presence of a period of negligible water velocity, known as slack water, exhibited by reversing currents. This interval occurs near the times of high and low water and is of importance in construction and other operations. Figure 4.6, reproduced from reference 1, shows the normal monthly variation of reversing currents for several U. S. ports. In river mouths the ebb flow is typically stronger and of longer duration than the flood.

Consider the case of an enclosed estuary open to the sea at one end by a narrow channel. During one-half of a tide period ($T/2$) a volume of water equal to the tide range (R_T) times the surface area of the basin (S) must flow through the channel of average cross-sectional area (A) (see Fig. 4.7). The average velocity through the channel will then be:

$$V_{avg} = \frac{2SR_T}{TA} \tag{4.2}$$

The maximum velocity across the channel will be $\pi/2$ times the average. Observations have shown that the velocity at mid-channel is about one-third greater than at the sides, thus the maximum current at mid-channel can be estimated from:

$$V_{\max} = \frac{4}{3} \pi \frac{SR_T}{TA} \tag{4.3}$$

Equation (4.3) can be used to estimate the maximum tidal current in an entrance channel; however, more precise calculations of tidal flow are very complex. The reader is referred to the work of Einstein[6] and the general discussion by Ippen and Harleman.[7]

A most interesting special case of tidal current, known as hydraulic current, occurs in channels connecting two tidal bodies of water that are subject to different ranges and times of high tide. Notable examples of this are the Cape Cod Canal in southern Massachusetts and the East River entering New York Harbor. The speed of hydraulic currents varies nearly as the square root of the difference in tide heights. Hydraulic currents typically reach maximum strength more quickly and exhibit shorter periods of slack water than normal tidal currents.[8,9]

It must be emphasized that in general it is best to rely upon published tide tables and current charts wherever possible and to make

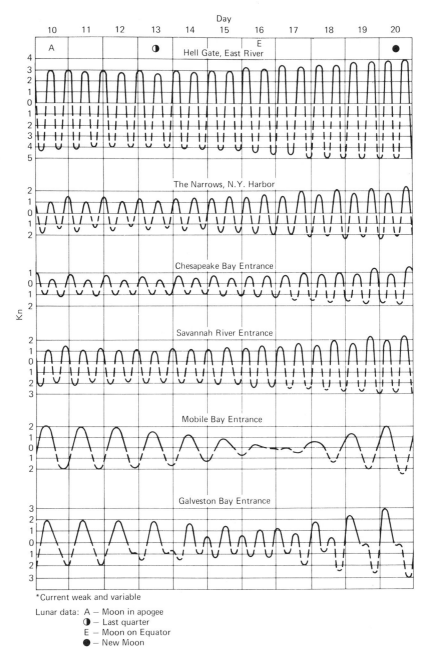

Fig. 4.6. Typical current curves for reference stations (Flood: solid line; ebb broken line). (From *Tidal Current Tables* NQAA., NOS.[1])

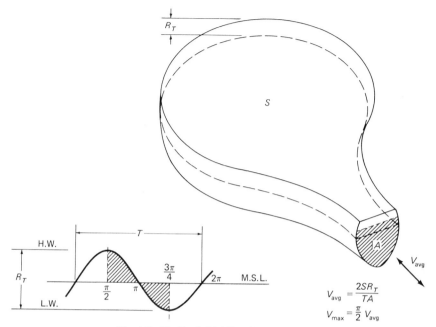

Fig. 4.7. Idealized tidal flow in enclosed bay.

$$V_{avg} = \frac{2SR_T}{TA}$$

$$V_{max} = \frac{\pi}{2} V_{avg}$$

use of actual field measurements when necessary rather than to attempt to calculate current velocities. There have been extensive current measurements taken from lightships off the U.S. coasts, notably those summarized by Marmer[10] for the Pacific Coast and by Haight[11] for the Atlantic Coast.

Wind-Stress Currents

Wind blowing over the ocean surface tends to drag the surface layers along with it via frictional shearing stresses. At higher wind velocities that are sustained for several hours or more, significant surface currents can be set up which are of particular relevance when added vectorially to the prevailing tidal currents (see Fig. 4.3). Surface wind-stress currents are directed to the right of the wind direction in the Northern Hemisphere (opposite in the Southern Hemisphere) and may attain velocities of around 1% to 5% of the sustained wind speed. Figure 4.8, reproduced from reference[12], in-dicates expected current velocities for various wind speeds and

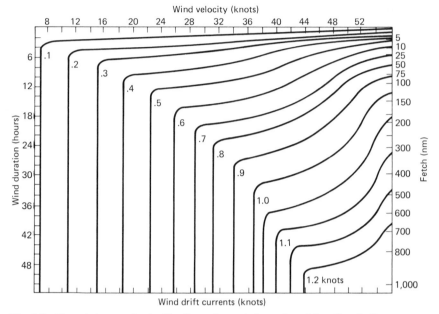

Fig. 4.8. Expected current velocities for various wind speeds and durations in the open sea. Enter with wind velocity and read drift current for the appropriate wind duration or fetch, whichever gives the smaller current. (From *Ocean Thermal Structure Forecasting*, U. S. Naval Oceanographic Office.[12])

durations in the open sea. It must be realized, however, that shallow depths and the proximity of land can greatly modify the current direction and speed with respect to the wind direction and speed. The tidal current tables produced by NOAA[1] give instructions for estimating wind-stress currents and include some observed values of relative direction and speed with respect to the wind direction and speed for a few typical reference stations. Table 4.1 is reproduced from the *Tidal Current Tables* and indicates observed velocities and directions of wind-stress currents relative to stated wind speeds and directions for 20 U. S. East Coast light stations.

Wind-stress currents diminish in strength rapidly and change direction with depth (Ekman's spiral), as previously discussed. The angle between the surface current and wind direction varies with latitude and wind velocity as well as with local bottom topography and the proximity of land. Note in Table 4.1 that some wind currents

Table 4.1. Wind-driven currents.

Average deviation of current to right or left of wind direction

[A minus sign (−) indicates that the current sets to the left of the wind]

Wind from ....

Old Lightship Stations	Lat. ° ′	Long. ° ′	N. °	NNE. °	NE. °	ENE. °	E. °	ESE. °	SE. °	SSE. °	S. °	SSW. °	SW. °	WSW. °	W. °	WNW. °	NW. °	NNW. °
Portland	43 32	70 06	24	14	9	8	−2	−14	0	26	15	18	18	24	15	34	13	18
Boston	42 20	70 45	---	−1	---	21	---	32	---	29	---	20	---	2	---	19	---	15
Pollock Rip Slue	41 37	69 54	6	5	48	−38	30	−53	−24	−75	−25	167	70	59	36	53	20	19
Nantucket Shoals	40 37	69 37	44	46	28	24	9	16	12	3	25	0	6	18	30	39	41	48
Hen and Chickens	41 27	71 01	16	14	−7	−1	−14	3	−39	−36	25	55	35	30	20	16	16	8
Brenton Reef	41 26	71 23	34	25	22	19	25	1	−7	8	27	48	23	41	41	31	21	24
Fire Island	40 29	73 11	35	23	15	8	2	−17	31	55	40	41	31	14	−2	0	25	37
Ambrose Channel	40 27	73 49	36	40	21	11	18	72	27	112	82	70	63	46	37	22	23	21
Scotland	40 27	73 55	16	−12	−26	−36	−61	−36	−92	−150	90	33	77	44	15	30	27	13
Barnegat	39 46	73 56	6	5	−13	−9	−16	−7	33	54	55	30	14	8	0	−5	21	29
Northeast End	38 58	74 30	30	14	−3	−11	−20	−31	−42	−28	37	44	25	18	7	16	25	18
Overfalls	38 48	75 01	28	−6	−1	2	−40	−56	−78	−22	68	28	55	54	32	31	32	45
Winter-Quarter Shoal	37 55	74 56	18	−1	−5	−21	−27	−35	−19	31	23	20	4	14	9	8	28	27
Chesapeake	36 59	75 42	18	−2	−4	5	−6	23	73	71	57	38	27	26	22	18	15	22
Diamond Shoal	35 05	75 20	11	3	−3	36	65	88	74	52	40	22	7	−10	−13	−17	−25	−4
Cape Lookout Shoals	34 18	76 24	30	24	2	2	−29	---	21	80	54	31	32	21	2	18	5	−5
Frying Pan Shoals	33 34	77 49	34	34	18	6	2	9	48	55	48	38	26	14	−7	−12	−27	−6
Savannah	31 57	80 40	12	12	−9	−18	−23	−46	17	50	43	17	7	−8	−10	7	15	33
Brunswick	31 00	81 10	17	−2	−10	−28	−18	−21	37	29	23	2	6	−21	−21	−26	16	18
St. Johns	30 23	81 18	3	−12	−27	−47	−84	30	35	26	26	27	1	16	−8	−17	6	8

Velocity. – The table below shows the average velocity of the current due to winds of various strengths.

Wind velocity (miles per hour)	10	20	30	40	50
Average current velocity (knots) due to wind at following lightship stations:					
Boston and Barnegat	0.1	0.1	0.2	0.3	0.3
Diamond Shoal and Cape Lookout Shoals .	0.5	0.6	0.7	0.8	1.0
All other locations	0.2	0.3	0.4	0.5	0.6

From *Tidal Current Tables*, NOAA, NOS. [1]

are actually driven to the left of the wind direction instead of the right as predicted by theory for ocean currents. Ekman's theory predicts an angle of 45° to the wind at the mid-latitudes.

After the wind or driving force stops, the current generated continues on as an inertial current until it is dissipated by friction. Inertial currents travel in elliptical paths varying from a hypothetical straight line at the equator to a circle at the North Pole, completing an orbit in time (T_I), where $T_I = 12$ hours/sin ϕ, and ϕ is the latitude in degrees.

Wind drift currents may be significant in structural design, especially when added to local tidal currents and/or wave particle velocities. Near shore and over the continental shelves currents will be stronger than normal during storms because of the associated abnormally high water levels (see Section 5.2). The velocity of these currents will be in approximate proportion to the increase in water level.

4.2 CURRENT DESIGN CRITERIA

Current design criteria should be selected in accordance with the importance of the structure and its sensitivity to current loads. Traditionally in the design of waterfront structures where current loads are not governing, current velocities are assumed and applied via semi-empirical formulas in a most simplistic way. In the design of large and important offshore platforms, however, current velocities and their spatial and temporal variations are determined by oceanographic and meteorologic specialists and are combined with other environmental loads in a more sophisticated manner. The design of most marine structures will usually fall somewhere between these extremes, and the designer must be familiar enough with the nuances of current behavior to determine the best approach to designing for current forces. Currents also have an important effect on the initial siting or orientation of a structure in terms of the structure's function; for example, berthing structures should be located such that berthing vessels are headed into the current direction as far as is possible.

The most important single factor in designing for current loads is the selection of the maximum design velocity. Selection of the maximum current velocity should be based on actual measurements when possible or on reliable current table data with consideration of seasonal or storm water level (see Section 5.2) maximum and the superposition of various interacting sources of currents (i.e., tidal, wind-stress, river-run-off, etc.). The vertical distribution, and in the case of rivers and narrow channels the horizontal distribution of current velocity, must also be considered. For the case of tidal currents in protected harbors acting on waterfront structures not sensitive to current loads it may be adequate to select a design velocity from tidal current table data or a few measurements and assume this velocity to be uniform with depth to simplify calculation. Selection of tidal current velocities must always consider the seasonal maximum spring tide (see Section 5.1) current, which is usually 10% to 20% stronger than the average tidal current.

In general, for shallow water sites where measurements are lacking the vertical distribution of tidal current velocity can be assumed to be distributed in accordance with the one-seventh-power law which gives the tidal current velocity at any depth (U_{Tz}) as:

$$U_{Tz} = U_{Ts}\left(\frac{z}{d}\right)^{1/7} \tag{4.4}$$

where U_{Ts} is the surface current, z is the distance above the bottom measured positive upward, and d is the water depth. Design guidelines for fixed offshore platforms[11, 12] suggest the use of this profile in the absence of specific site data. The tidal current profile should be added vectorially to the maximum wind-stress current, which can conservatively be assumed to be linearly distributed from zero at the mudline to a maximum at the surface. In reality, wind-stress currents are usually only of significant velocity near the surface. The vector addition of these vertical profiles is illustrated in Fig. 4.9. As strong wind-stress currents and higher-than-normal water levels are generally associated with high waves, the current velocities must be added vectorially to the wave particle velocities (Section 3.6) and integrated

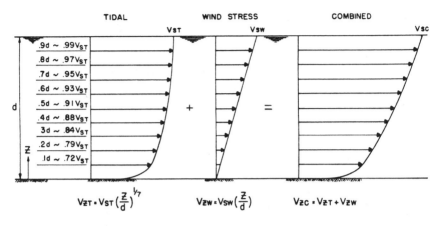

$$V_{\bar{z}T} = V_{ST}\left(\frac{z}{d}\right)^{1/7}$$ $$V_{\bar{z}W} = V_{SW}\left(\frac{z}{d}\right)$$ $$V_{\bar{z}C} = V_{\bar{z}T} + V_{\bar{z}W}$$

OFFSHORE FOR SMOOTH, HORIZONTAL BOTTOM THE ABOVE
PROFILES CAN BE USED WHEN ACTUAL MEASUREMENTS ARE
NOT AVAILABLE.

Fig. 4.9. Current velocity profiles assumed for structure design. (From reference[17])

over the water depth to find the maximum total force and overturning moment on a structure.

For deep water structures the current velocity profile as well as
seasonal maximums and averages should be determined by oceanographic specialists and/or a comprehensive measurement program.

Hall[13] describes such a program and demonstrates the significance
of hurricane-generated ocean currents to the design of offshore
structures in the Gulf of Mexico. In an example of the force calculation for a 3.7-foot-diameter pile situated in 300 feet of water
and subjected to a 50-foot, 12-second wave, Hall indicates the combined wave and current force is nearly 70% larger than the force
from the wave alone, although in his example the surface current
velocity is only 4.1 feet per second compared to a maximum wave
particle velocity of about 15 feet per second for the design wave.

Wind-stress currents may also be significant in the along-shore
direction. Bretschneider[14] developed the bathystrophic technique
of predicting such longshore currents. Some further discussion of
currents associated with storm surge is given in Section 5.2. Other
theoretical hindcast methods of predicting storm-generated currents
for design purposes are described by Goldman[15] and by Welander.[16]

In narrow channels the distribution of velocities across the channel
as well as with depth should be considered. Hypothetical velocity
contours for a straight uniform channel are shown in Fig. 4.10. In

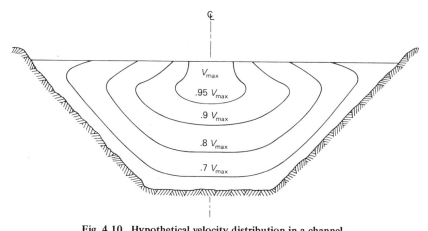

Fig. 4.10. Hypothetical velocity distribution in a channel.

constricted areas the presence of the structure itself may modify the flow of water, thus increasing local velocities. The pattern of flow thus set up can only be guessed at without hydrodynamic modeling. Floating structures moored near a bank may be subjected to suction forces away from the bank or sinkage forces, due to the increased velocity around one side or below the structure, respectively.

4.3 CURRENT FORCES

Once design current velocities and the appropriate vertical distribution have been selected, the current loads can be evaluated using the drag force equation (equation 2.5), introduced in Section 2.3. Drag, lift, and possible dynamic amplification of forces should be considered.   Lift forces (hydrodynamic force acting perpendicular to the direction of flow) may be of critical importance in the case of long slender cylindrical components, such as pipelines with long unsupported lengths, and may be responsible for exciting harmonic motions such as discussed in Section 2.4.   The drag force (acting in the direction of flow) is generally of the most concern in the design of fixed and floating structures.   For exposed structures where wave forces are important, the current velocities should be added vectorially to the wave particle velocities (u) to yield the total drag on the structure (F_{TD}) where:

$$F_{TD} = \frac{1}{2} \frac{\gamma}{g} C_D A_P (U + u)^2 \qquad (4.5)$$

and γ is the fluid density, approximately 64 lb/cu ft for seawater, g is the acceleration of gravity, A_p is the projected area, C_D is the drag coefficient, which may be selected from Table 2.3 and Fig. 2.7 through 2.9, and U and u are the current and wave particle velocity, respectively. The above equation must be integrated vertically over the velocity profile to yield the maximum total force and overturning moment on the structure. In selecting drag coefficients, consideration should be given to the effects of various types and thicknesses of fouling (see Section 7.3).

For the case of more traditional waterfront structures where current forces are not as critical, the current pressure for the mean or maximum current can be assumed uniformly distributed over the entire structure, and the curves in Fig. 4.11, reproduced from reference 17, may be useful in estimating the current drag force for various drag coefficients. For flowing seawater and for the current velocity (U) in feet per second and the force (F_c) per unit area (A_p) in pounds per square foot, equation (2.5) reduces to:

$$\frac{F_c}{A_p} = U^2 C_D \qquad (4.6)$$

Therefore, for a current velocity of 4 fps (= 2.4 knots) the dynamic pressure is $(4)^2 = 16$ psf, which is obtained from Fig. 4.11 for the curve of $C_D = 1.0$. The dynamic pressure then need only be multiplied by an appropriate drag coefficient for the given structure times its projected area.

Special consideration should be given to conditions such as narrow channels and river mouths where the construction of a new structure may alter the flow of water. Reference 18, for example, indicates that in the case of floating structures with water depth (d) to draft (D) ratios (d/D) less than 6, the drag coefficient (C_D) rises exponentially to a figure at $d/D = 1$ approximately six times the value that it was at $d/D = 6$ or greater. Where important structures are to be built in constricted areas, model testing is the only reliable way to predict the modified flow and resulting force distribution. Figure 4.12, reproduced from reference 19, shows the variation in lateral force with water depth on a moored Liberty ship based upon model tests conducted at the David Taylor Model Basin. This figure can be used to estimate the drag force on other-size vessels of similar configuration by applying the laws of mechanical similitude as described in Section 2.3 for wind forces on the same vessel. The linear ratio in this figure is 48.6, and the values of lateral resistance are for a 4-knot current, which can be scaled to other current velocities in proportion to the squares of the velocities. Figure 4.12 illustrates the rapid increase in drag force with de-

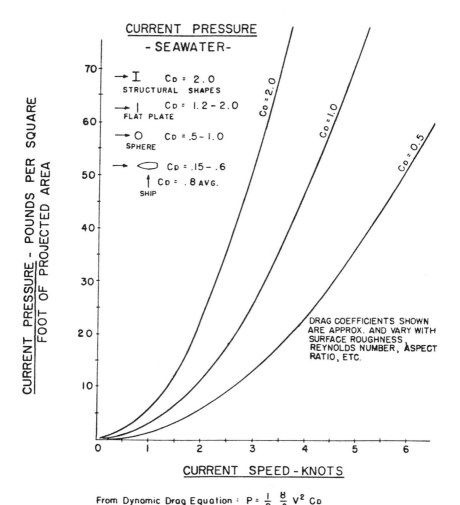

Fig. 4.11. Current pressure—seawater. (After Gaythwaite[17])

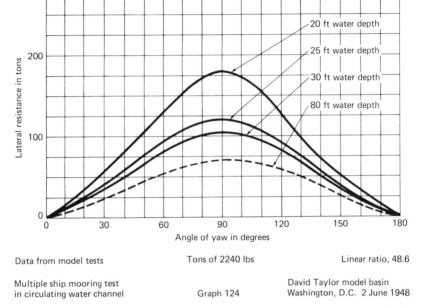

Data from model tests   Tons of 2240 lbs   Linear ratio, 48.6

Multiple ship mooring test   David Taylor model basin
in circulating water channel   Graph 124   Washington, D.C. 2 June 1948

Fig. 4.12. Variations in lateral resistance with depth of water of a single moored liberty ship of the EC-2 class in a 4-knot current. (After U. S. Navy.[19])

creasing water depth.   The prototype vessel has a mean draft of about 10 feet, so the upper and lower curves correspond to d/D ratios of approximately 2 and 8, respectively, and demonstrate an approximately 260% increase in lateral force for the shallower water.   Figure 4.13, also from reference 19, shows the longitudinal force and yawing moment as well as the lateral force on the same vessel represented in Fig. 4.12, in a water depth of 25 feet, for a 4-knot current.

From Fig. 4.13 it can be seen that the maximum lateral force on the vessel with the current abeam is approximately 120 long tons (1 long ton = 2240 lb), or 268,800 lb.   Rearranging equation (4.5):

$$C_D = \frac{F_C}{U^2 A_P} = \frac{268,800 \text{ lb}}{(4 \text{ kt} \times 1.69 \text{ fps/kt})^2 \times 410 \text{ ft} \times 10 \text{ ft}} = 1.43$$

where the vessel's waterline length is 410 feet and the draft used in the model test was 10 feet (in this case $d/D = 25'/10' = 2.5$).   This calculation is only approximate, since the projected area of the hull is actually less than 410′ × 10′ owing to the vessel's underwater configuration, but it serves to illustrate the variations in drag coefficients. Referring to Fig. 4.12, the same vessel in a water depth of 80 feet has approximately 58% the lateral resistance to what it is in 25 feet of water; therefore, the effective drag force coefficient in deep water would be about C_D = .58 × 1.43 = 0.83. We calculate

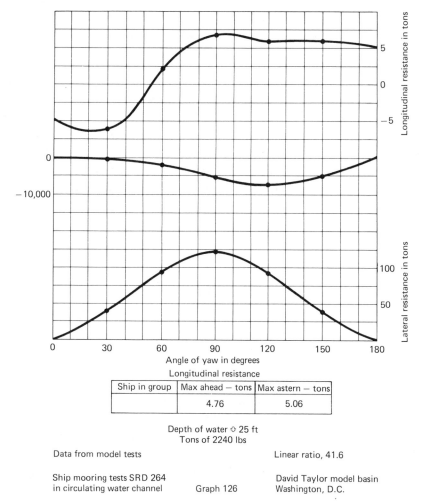

Ship in group	Max ahead — tons	Max astern — tons
	4.76	5.06

Depth of water ⟡ 25 ft
Tons of 2240 lbs

Data from model tests

Linear ratio, 41.6

Ship mooring tests SRD 264
in circulating water channel

Graph 126

David Taylor model basin
Washington, D.C.

Fig. 4.13. Forces acting on one moored vessel of EC-2 class in 4-knot current. (After U. S. Navy.[19])

similarly for the longitudinal force, which from Fig. 4.13 is about 5 tons maximum, = 11,200 lb, with the vessel's beam at the waterline (from reference 19) 57 feet:

$$C_D = \frac{11,200 \text{ lb}}{(4 \times 1.69)^2 \times 57' \times 10'} = .43$$

which in deeper water reduces to $C_D = .25$.

Other useful curves for various vessel types and configurations can be found in reference 19. Reference 20 contains valuable model test data for predicting wind and current loads on very large crude carriers (VLCC's), which would be most useful in the design of offshore terminals and sea islands. These tests were conducted at the Netherlands Model Ship Basin for vessels in the size range of 150,000 to 500,000 DWT.

In the case of moored vessels and floating structures that are relatively long in the direction of flow, it is customary to add skin friction resistance (F_{SF}) to the drag force, although for many shapes this can be accounted for in the selection of an overall drag coefficient. Skin friction varies with surface roughness and Reynolds

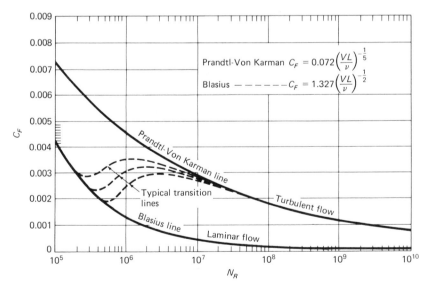

Fig. 4.14. Skin friction lines, turbulent and laminar flow. (From Comstock.[21])

number (N_R). Figure 4.14, from reference 21, shows the variation in the skin (or surface) friction coefficient (C_F) with N_R for various flow regimes. For new structures with smooth surfaces and fresh-water structures not subject to fouling, C_F can be taken as .001 in most cases. In marine structures subjected to hard fouling (see Section 7.3) the effect is to shift from laminar toward a more turbulent flow, and a C_F of .004 to .006 should be used. Surface friction varies with the square of the velocity (U) and acts over the wetted surface area (S) as expressed by the equation:

$$F_{SF} = \frac{1}{2}\ \frac{\gamma}{g}\ SC_F\ U^2 \qquad (4.7)$$

It is also customary to account for special appendages such as a vessel's propellers with a separate drag force term. Propeller drag can be estimated using the drag force equation times a drag coefficient of 1.0 applied to the total area circumscribed by the tips of the propeller blade.

At the air-sea interface, free surface waves are produced by bodies traveling through the water. Wave-making resistance is all important in the design of vessel hulls and power plants, but generally unimportant in the design of stationary structures because of the low speed-length ratios involved. The speed–length ratio is the velocity (V) divided by the square root of the vessel's waterline length (L_{WL}). For V in knots and L_{WL} in feet a speed length ratio of $V/\sqrt{L_{WL}} = 1.34$ corresponds to the celerity of a free surface wave, which means that the vessel must "climb" up over the wave crest formed at its leading edge, which in turn requires tremendous relative amounts of propulsive energy. A vessel that exceeds a $V/\sqrt{L_{WL}} = 1.34$ is mathematically planing. Wave-making resistance is generally found by model testing and is discussed in most naval architecture texts. It is rarely important in structural design and is beyond the scope of this text. When vessels or floating structures are to be moored in very strong currents, it should be investigated. Another way to find the total resistance of vessels is to use powering and resistance formulas or curves found in naval architecture texts such as reference 21, or to conduct model tests or compare the vessel or structure with model tests results for similar shapes and apply the laws of similitude as previously described.

In shallow water, strong currents may cause lateral vibration problems due to vortex shedding and structure resonance. Khanna and Wood[22] describe such a problem and its solution during the construction of an ore terminal in 6-knot tidal currents. In this case, free-standing piles awaiting the deck structure would have been subjected to severe oscillations; so a tripod solution was found based upon model tests and theoretical considerations.

Currents may cause impact forces by carrying ice or debris into a structure. Impact forces in general can be evaluated by relating the kinetic energy of impact to some deformation of the structure if the mass of the impacting body is known. This is a particularly important consideration in areas subject to moving icebergs, which can be of considerable size and travel at the speed of the current (see Section 6.3).

4.4 OTHER EFFECTS OF CURRENTS

General Considerations

In addition to being a source of hydrodynamic loads, currents may affect structures in many other ways. By serving as a transport medium, currents may carry flotsam (floating debris) or ice into the structure causing impact forces, scouring and deposition at the base of the structure, abrasion of the structure via entrained sand, for example, and the introduction of fouling or boring organisms (see Sections 7.3 and 7.4) borne by shifting currents. Because the rate of corrosion of metals in seawater is directly proportional to the water velocity, currents usually increase corrosion rates, a consideration discussed in some detail in Section 7.5. Currents modify the properties of waves, as discussed in Section 3.3. Currents may have a profound effect on siting of structures, on littoral drift problems, and on operations, all of which are beyond the scope of this text.

Scouring and Deposition

Sands and silts are particularly susceptible to erosion by moving water. Figure 4.15 indicates threshold velocities at which given sand particle sizes may be picked up and carried by currents. When the

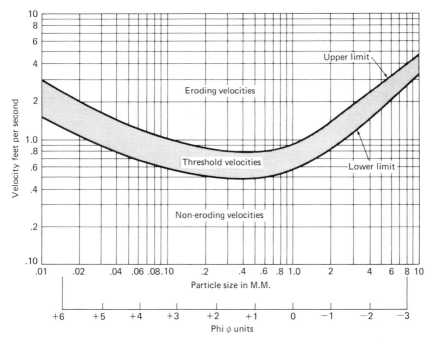

Fig. 4.15. Mean velocities required to erode sand. (After U. S. Army, CERC, SPM.[5])

current velocity is reduced, the entrained sand will be deposited. Scouring is an important consideration in structural design, as the removal of sand may cause undermining of mat foundations and increase the effective column length of pile type structures. Reference 23 indicates some preferred geometries for mat type foundations. In general, mat foundations should be designed with adequate scour protection or with the assumption that 20% to 50% of the bottom support will be removed by scouring. Likewise, the design of pile structures should consider the effects of increased unsupported length of piles.

REFERENCES

1. *Tidal Current Tables*, 2 vol. covering the Atlantic and Pacific coasts of North America and Asia, U. S. Dept. of Commerce, NOAA, published annually.

2. Ekman, V. W. "On the Influence of the Earth's Rotation on Ocean Currents," *Ark. Mat. Astron. Fys.* Stockholm, 2, 11, 1905.

3. Rossby, C. G., and Montgomery, R. B. "The Layer of Frictional Influence in Wind and Ocean Currents," papers in phys. oceanog., MIT and WHO I, 3, April 1935.

4. Shepard, F. P. *The Earth Beneath the Sea*, (pp. 23-26), Atheneum, New York, 1969.

5. U. S. Army, Coastal Engineering Research Center. *Shore Protection Manual*, 3 vols., USACE, 1977.

6. Einstein, H. A. "Computations of Tides and Tidal Currents – United States Practice," *Proc. ASCE*, 81 (715), 1955.

7. Ippen, A. T., and Harleman, D. R. F. "Tidal Dynamics in Estuaries," *Estuary and Coastline Hydrodynamics*, McGraw-Hill, New York, 1966.

8. Brown, "Flow of Water in Tidal Canals," *Trans. ASCE*, Vol. 96, 1932.

9. Wilcox, "Tidal Movement in the Cape Cod Canal," *Proc. ASCE, Journal Hydraulics Div.*, April 1958.

10. Marmer, H.A. *Coastal Currents Along the Pacific Coast of the United States*, USCGS, S.P.–121, 1926.

11. Haight, F.J. *Coastal Currents Along the Atlantic Coast of the United States*, USCGS, S.P.–230, 1942.

12. U. S. Navy. *Ocean Thermal Structure Forecasting*, USNOO, ASWEPS Manual Series, Vol. 5, SP–105, 1966.

13. Hall, J.M. "Hurricane-Generated Ocean Currents – Part I: The Development of a Measurement Program," *OTC* paper No. 1518, Houston, May 1972.

14. Bretschneider, C. L. "On Wind, Tide, and Longshore Currents Over the Continental Shelf Due to Winds, Blowing at an Angle to the Coast," National Eng'g. Science Company, Report to Office of Naval Research, Washington, D.C., Dec. 1966.

15. Goldman, J. L. "The Effect of Time Dependent Transports in Storm Generated Currents," *OTC* paper #1347, Houston, April 1971.

16. Welander, P. "Numerical Prediction of Storm Surge," *Advances in Geophysics*, Vol. 8, Academic Press, New York, 1961.

17. Gaythwaite, J.W. "Structural Design Considerations in the Marine Environment," *Journal, Boston Society Section, ASCE*, Vol. 65, No. 3, Oct. 1978.

18. Evans, H. and Adamchak, J. *Ocean Engineering Structures*, M.I.T. Press, Cambridge, Mass., 1969.

19. U. S. Navy. *Design Manual – Harbor & Coastal Facilities*, NAVFAC, DM-26, Washington, D.C., July 1968.

20. Oil Companies International Marine Forum. *Prediction of Wind and Current Loads on VLCC's*, London, 1977.

21. Comstock, J. P., ed. *Principles of Naval Architecture* SNAME, New York, 1967.

22. Khanna, J., and Wood, J.S. "Tripod Structures in Fast Currents," *The Dock and Harbour Authority*, Vol. LIX, No. 700, March, 1979.

23. Ninomiya, K. "A Study on Suction and Scouring of Sit-on-Bottom Type Offshore Structure," *OTC*, paper #1605, Houston, May 1972.

5
Water Level Variations
and Long Wave Effects

Variations in water levels, in particular the maximum and minimum levels that will occur at a site, may be of profound importance in the design of fixed and floating structures. Aside from affecting siting and operations, the maximum water level affects the height of a fixed platform and the scope of mooring lines of floating structures, minimum water levels may affect the stability of bulkheads, and the total range in water level affects the vertical extent to which the structure is affected by degrees of corrosion and fouling, and so fourth. Water level fluctuations, discussed in this chapter, may vary over a period of minutes, hours, or years, as distinguished from the instantaneous variations associated with the passage of wave crests. Most of the mechanisms responsible for long-term water level variations can be considered a form of long waves: the astronomic tides, for example, with periods of 12 to 24 hours and wavelengths equal to half the circumference of the earth; the storm-related extreme tides or surges, which are composed of several components of meteorological origin with periods ranging from minutes to hours; seismically generated waves called tsunamis; and resonant oscillations in enclosed and semi-enclosed basins called seiches, again with periods in the range of minutes to hours. There are also seasonal, annual, and longer-term geological and climatological effects.

The above-mentioned phenomena will be discussed herein with particular regard to the establishment of maximum design water levels to which maximum wave crest elevations must be added, and minimum water levels which may determine wave steepness and breaking characteristics. In this regard the astronomic tide and so-called storm tide are of the most concern; thus these phenomena are discussed in some-

what more detail than seiches and tsunamis, which are more important from operational – and, in the case of tsunamis, disaster prevention– points of view.

5.1 THE TIDES

General Discussion

Tides are the periodic rise and fall of sea level in response to the
gravitational attraction of the sun and moon as modified by the earth's
rotation, friction forces, and the ocean boundaries. Tidal theories
are treated in detail in most physical oceanography texts such as the
general oceanography references given at the end of Chapter 1. Even
the most sophisticated theories are far from exact because of the
many interacting periods associated with the subtle movements of
the earth, sun, and moon, the complex geometry of the ocean basins,
and the fact that simplified two-dimensional analysis cannot be applied because coriolis force and other inertial accelerations cannot be
neglected. The equilibrium theory of the tides was one of the earliest
theories and is useful for illustrating the overall behavior of the liquid
oceans on a solid earth. In accordance with Newton's law of universal gravitation (i.e., every body attracts every other body with a force
directly proportional to its mass and inversely proportional to the
square of the distance between the bodies) the sun and moon exert a
pull on the earth's surface. The net tide-producing force, however, is
inversely proportional to the cube of the distance when we consider
the effect of gravity at the earth's surface. The sun, being much more
distant, exerts only about 46% the force of the moon; hence, the tides
generally follow the movements of the moon more than those of the
sun. In very simple terms, the net effect is that the acceleration of
gravity at the earth's surface directly beneath the moon is effectively
reduced because of the moon's attractive force, and water is thus
heaped up about this sub-lunar point. Because the earth and moon
rotate about a common center of mass, which is located inside the
earth, a centrifugal force is produced that, combined with the effect
of the lesser attraction of the moon on the far side of the earth,
causes the earth's own surface gravity to be effectively reduced by
the same amount as the side facing the moon. Figure 5.1 shows the

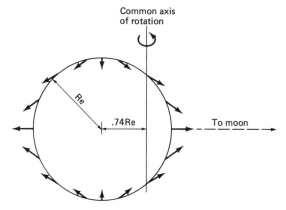

Fig. 5.1. Resultant tide ("tractive") forces.

resulting tide-producing or tractive forces. The earth's rotation beneath these tidal bulges causes the typical (semi-diurnal) effect of
two high and two low tides a day.

One of the earliest attempts to develop a coherent theory of the
tides and the ability to predict them is known as the equilibrium
theory, which is far from being exact but is useful for instructional
purposes. Sir Isaac Newton assumed the earth to be covered uniformly
by an ideal ocean of equal depth everywhere in order to simplify
tidal analysis. He encountered difficulty, however, when he realized
that the tidal bulges could not follow the moon's motions exactly
because of friction forces and coriolis deflection.

The dynamical theory of the tides as proposed by Laplace, an im-
provement upon Newton's equilibrium theory, attempts to relate the
hydrodynamic response of the oceans to the astronomical forcing
functions, but because of the aforementioned complexities it is not
satisfactory for tide prediction. Tide prediction is presently based
upon a so-called harmonic analysis whereby the movements of the
sun and moon are broken into various periods during which some
cycle of motion is repeated. U.S. tide prediction machines use about
37 harmonic components, while some European machines use 60 or
more. Doodson[1] listed some 390 components of tide-generating potential. A few of the more important variations will be discussed here.

The moon goes through characteristic phases, the major ones being
full moon, first quarter, new moon, and last quarter, in a period of

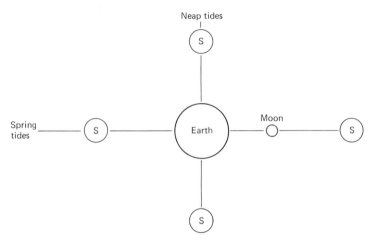

Fig. 5.2. Relative position of sun at springs and neaps.

29.5306 days called a synodic month. When the moon is full or new, it is in line with the sun, and their attractive forces are additive, so that higher-than-normal (spring) tides result (see Fig. 5.2). Conversely, at first quarter and last quarter the moon is at quadrature with the sun, their effects are subtractive, and lower-than-normal (neap) tides result. Tides at springs and neaps are usually 10% to 20% greater than or less than the mean range, respectively. The moon's orbital period, called a sidereal month, is shorter than the synodic month owing to the earth's motion about the sun. While the earth rotates once, the moon makes 1/29.5 of a revolution in the same direction of rotation so that the lunar period, designated M_2 in harmonic analysis, is equal to 12 (1 + 1/29.5) or 12.41 hours. The principal solar period (S_2) is 12.0 hours, but because the moon predominates, the semi-diurnal tide progresses about .82 hour or 49 minutes each day.

Both the sun and the moon move north and south of the plane of the earth's equator, and during the time they are at the equator higher tides (the equinoctial tides) result (see Fig. 5.3). The highest spring tides are usually noted in the early spring and early fall as the sun crosses the equator. The moon's extreme declination varies between 18 1/2° and 28 1/2° north and south over a period of about 18.6 years (the nodal period); so all records should be averaged over this length of time to obtain an accurate average. The moon travels

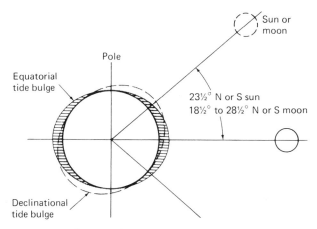

Pole

Equatorial tide bulge

Sun or moon

23½° N or S sun
18½° to 28½° N or S moon

Declinational tide bulge

Fig. 5.3. Effect of declination on tides.

between its extreme north and south declinations in one tropical month of 27.3217 days. Lower-than-usual (tropical) tides occur when the moon is at its extreme declination.

Because the moon's orbit about the earth is elliptical, there are times when it is closest to the earth (perigee) and its influence is greatest and times when it is farthest from the earth (apogee), when its influence is least. The period between apogee and perigee is termed an anomalistic month and is equal to 27.5546 days. Perigean and apogean tides are approximately 20% greater than or less than normal tides, respectively. Similarly, once a year the earth is closest to the sun (perihelion) and once a year is farthest from the sun (aphelion). The tide-producing effects of some of the major variations in the motions of the moon and sun are shown in Table 5.1, which lists the principal tidal harmonic components used in tide prediction. These major components and many lesser ones are all superposed to give the tide-generating potential at any given time. For example, at least twice a year exceptionally high, perigean spring tides occur. The effect of the planets is negligible on the earth's tides.

The tide prediction problem is still more complicated because of the complex geometry of the oceans and its attached basins. Semi-enclosed and enclosed basins tend to oscillate at some natural frequency or even multiples of that frequency (see Section 5.4), so that when its natural period is close to the tide-producing period, the lo-

Table 5.1. Principal tidal harmonic components.

Description	Symbol	Period (hours)	Magnitude ratio $(M_2 = 100)$
Semi-diurnal components:			
Principal lunar	M_2	12.42	100.0
Principal solar	S_2	12.00	46.6
Lunar monthly	N_2	12.66	19.1
Soli-lunar declinational	K_2	11.97	12.7
Diurnal components:			
Soli-lunar	K_1	23.93	58.4
Principal lunar	O_1	25.82	41.5
Principal solar	P_1	24.07	19.3
Long period components:			
Lunar fortnightly	M_f	327.86	17.2

cal tide will be amplified. This is most noticeable in elongated basins such as the Bay of Fundy in Nova Scotia and New Brunswick where 30- to 40-foot tide ranges are observed. Because tides propagate as a wave phenomenon at approximately the long wave speed $(C = \sqrt{gd})$ and are subject to the same physical laws as surface gravity waves, they are also slowed by bottom friction. In the upper reaches of the Bay of Fundy in the Petitcodiac River, the crest of the incoming tide wave travels at a faster rate than the trough of the previous low tide, because the speed of a tide wave is determined by the water depth. The higher water overtakes the retarded low water trough, and it charges up the river as a virtual wall of water several feet high, known as a bore. Spectacular bores also exist in the Amazon River (up to 25 feet high!), and in the Chien-Tang River of China. Bores are an interesting but relatively rare phenomenon not generally of interest in structure design, but they are discussed further in Section 5.5 in connection with tsunami design considerations.

Tides propagate everywhere at sea at the long wave speed (see Section 3.1) because the length of the tide wave is everywhere very long compared to the shallow ocean basins. At sea and at oceanic islands, the tidal amplitude is very small, perhaps one foot or less. It is only in the relatively shallow water such as that found over the continental shelves that the amplitude begins to build.

Along open coasts there is a more-or-less uniform or a uniform progression of tide heights and times; but within bays and estuaries rather drastic local effects may be observed because of the restrictive topography and possible resonant effects (see Section 5.4). As at oceanic islands, tides within relatively landlocked areas such as the Mediterranean Sea are quite small (on the order of one foot or less); in the Great Lakes astronomical tides are negligible; along open coastlines over the continental shelves tides are a few feet high; where as within gulfs, bays, and estuaries they may be several feet to several tens of feet high. In the upper regions of the Bay of Fundy, for example, 40-foot tides are usual, and tides of up to 56 feet have been recorded! Tidal waves appear to propagate outward from imaginary amphidromic points, at which there is virtually no change in water elevation and which serve as centers of rotation for the radiating tide waves. Such rotational waves are known as Kelvin waves. Co-tidal lines, which radiate outward from the nodal points, can be plotted on charts which serve to illustrate locations of equal time of tide. Co-range lines can also be plotted to show locations of equal tide range.

At some places on earth, the period of oscillation of the local basin may conflict with the normal twice-a-day semi-diurnal tide and result in diurnal inequalities (i.e., one of the day's tides will be higher or lower than the previous tide. At some locations, which respond to the diurnal components, one of the day's tides will be eradicated, resulting in a diurnal tide of one high and one low per day. Tides falling in between are called mixed tides. Figure 5.4 has been excerpted from the *Tide Tables,*[2] published annually by The National Oceanic and Atmospheric Administration (NOAA), National Ocean Survey (NOS), which shows typical tide curves for several U.S. ports. The engineer must rely on published tide table data and observations because of the complexities involved in prediction. Tide heights and times for remote locations can be estimated. Reference 3 gives instructions for estimating tide heights using precomputed values of the M_2, S_2, K_2, O_1, and other components. The degree of diurnal inequality depends upon the ratio of $K_1 + O_1$ to $M_2 + S_2$. When this ratio is around 0 to .25 tides will be semi-diurnal; when it is above .25 up to 3.0, they will be mixed, with the semi-diurnal and diurnal components dominating in the lower and higher ends of the range, respectively. When the ratio is greater than 3.0, the tide will be purely diurnal. Defant[4] demonstrates the application of

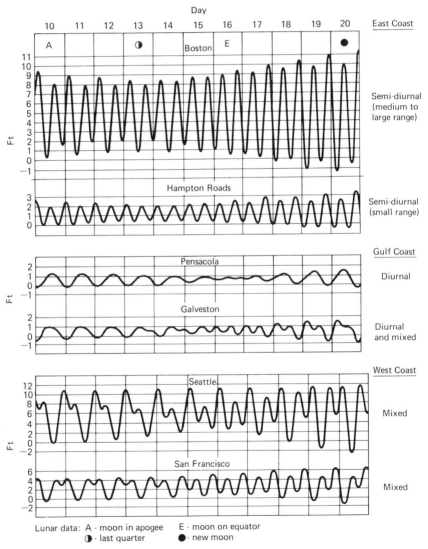

Fig. 5.4. Typical tide curves for selected U.S. ports. (From: "Tide Tables", NOAA, NOS[2])

the above to some typical world ports and lists values of the principal tide-producing components for various locations. Reference 5 indicates the types of tides found within the North Atlantic Ocean and also gives co-tidal lines and other valuable tidal information for the North Atlantic.

Because of friction and the earth's rotation, high tides may lag or precede the passage of the moon by a few hours, phenomena known as lagging and priming of the tides, respectively. The period between meridian passage and high water is called the luni-tidal interval. The "establishment of the port" or just the "establishment" refers to the lag between a full moon meridian passage and the highest spring tide, which may be up to 2 days. The effect of the earth's rotation is to deflect the tide wave to the right in the Northern Hemisphere (and to the left in the Southern Hemisphere), so that in bays and channels tides tend to occur sooner and are higher on the right side than on the left.

Because tide waves are of low steepness, they exhibit a high reflectivity, which increases the complexity of the tide in some regions.

Tidal Datums

The construction of all structures requires the establishment of some vertical reference plane, and it is most important that the engineer understand the chart and survey datum in use in the locality in which he is working and the relation of changes in water level to that datum. Mean sea level (M.S.L.) can be defined as that average sea level about which the tides oscillate. As stated previously, it must be determined over a period of record of about 19 years.

The current period of record, termed the National Tidal Datum Epoch, which has been adopted by the NOS, is the period 1941 through 1959. At certain locations M.S.L. and the mean tide level (M.T.L.), half the distance between mean low water (M.L.W.) and mean high water (M.H.W.), may differ slightly. In certain harbors the datum or benchmark in present use may not coincide exactly with M.S.L. or M.L.W. as currently employed by NOS. It is acceptable to use such a datum to be consistent with local municipal or navigation purposes, but one should be aware of its relation to normal and extreme variations in water levels. The term astronomical tide is used to distinguish the normal variations in water level caused by the sun and moon and their interactions from the so-called storm tide, which is caused by local meteorological effects. M.S.L. is important in establishing a fundamental benchmark for land survey-ing and is the basic vertical control for the U.S. precise level net.[6]

In 1929, the U.S. Coast and Geodetic Survey (USCGS) (now part-ly absorbed by NOAA) revised all nationwide leveling benchmarks to suit the mean sea level determined at that time. This is termed the USCGS mean sea level datum of 1929, and is currently the datum used in most references to M.S.L. In most parts of the United States, however, indications are that sea level has been rising relative to the land and at various rates for different locations. These long-term changes are relatively small and usually can be neglected for most engineering applications (see Section 5.3). M.S.L., however, is not generally used as a reference plane for marine structures. Nautical charts in the United States use mean low water (M.L.W.), which is the average of all low tides at a location or mean lower low water (M.L.L.W.) in areas subject to diurnal inequality, which is the average of the lowest low water of each tidal day. In Canada the lowest normal low water (L.N.L.W.) is used, which is the average of the monthly lowest astronomical tides. This practice is of obvious impor-tance to navigational interest. Thus, although in different parts of the world other datums may be used, it has been determined by in-ternational agreement that the chart datum shall be some level below which tides infrequently fall.

Figure 5.5 illustrates the relation of some tidal reference levels. Reference 7 provides a thorough discussion of tidal datums oriented toward navigational purposes, and Marmer[8] has summarized tidal da-tum planes in detail. Table 5.2 lists some typical tide ranges for selected North American coastal locations. Tide tables usually list the ranges of tide at many subordinate stations as well as the time and heights of tides for the primary reference stations. Tide tables are published annually for the U.S. coasts and other world locations by NOAA, National Ocean Survey.[2] Summaries of tidal observations around the United States may be found in Disney[9] and Harris and Lindsay.[10] Up-to-date tide information for a given location may be obtained upon request from NOAA and/or from the U.S. Army Corps of Engineers (USACE).

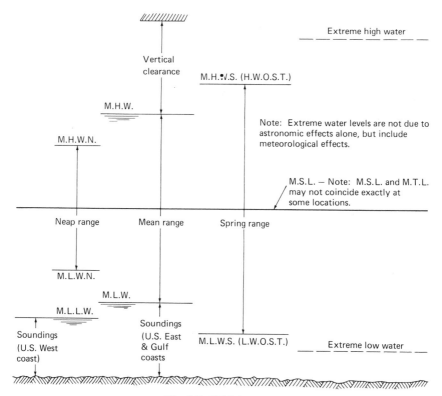

Fig. 5.5. Tidal datums.

5.2 STORM TIDE

General Description

Coastlines may be subject to higher (and lower) than normal water levels associated with meteorological conditions such as strong winds and low atmospheric pressures, occurring during periods of intense storm activity and known as storm tide or surge. On coasts with shallow water depths and those located where the continental shelf is wide and in enclosed bays, in estuaries, and on lakes, this pile-up or draw-down of water level may be considerable, on the order of several feet or more.

One of the most notable and tragic examples of storm surge occurred at Galveston, Texas in 1900. Hurricane winds of over 100

Table 5.2. Tide ranges for selected North American locations.

Location	Mean range (ft)	Spring range (ft)	Type of tide	Chart datum
Quebec, P.Q.	13.7	15.5	mixed	L.N.L.W.
St. John, N.F.	2.6	3.5	semi-diurnal	
Halifax, N.S.	4.4	5.3		
St. John, N.B.	20.2	23.6		
Portland, Maine	8.9	10.2		M.L.W.
Boston, Mass.	9.5	11.0		
New York, N.Y.	4.4	5.3		
Philadelphia, Pa.	5.9	6.2		
Baltimore, Md.	1.1	1.3		
Norfolk, Va.	2.8	3.4		
Charleston, S.C.	5.1	6.0		
Jacksonville, Fla.	2.0	2.3		
Miami, Fla.	2.5	3.0		
Galveston, Tex.	1.0	1.4	mixed	
San Diego, Calif.	4.2	5.8		M.L.L.W.
San Francisco, Calif.	4.0	5.7		
Columbia River, Oreg.	1.8	2.4		
Seattle, Wash.	7.6	11.3		
Juneau, Alaska	14.0	16.6		
Anchorage, Alaska	26.7	29.6		

Compiled from *Tide Tables*, NOAA, NOS. [2]

knots caused an increase in water level of 15 feet above the usual 2-foot tide range, allowing large breakers to destroy the city and drown 5000 people.[11]

Since, and prior to, that time numerous other events have caused extensive damage and loss of life. In the winter of 1953, a North Sea gale (later termed a 400-year storm), piled up water 10 feet above the highest tide level on the Dutch coast, overtopping and breaching dikes, resulting in nearly 1800 deaths and over $250 million in property damage. Progress in surge prediction and the use of protective structures have helped to minimize such loss in this country.

Reference 12 describes hurricane flood protection measures in the United States and concludes that in many locations construction of protective structures can be economically justified even though extreme water levels may be associated with long return periods.

In the design of marine structures in general, the determination of a maximum and sometimes a minimum design water level (D.W.L.) is of the utmost importance. Deck elevations must be above the maximum water elevation plus the crest elevation of the highest wave expected at the site. Offshore structures such as fixed oil platforms generally have an air gap of several feet above the crest of the highest expected wave. The increase in water depth due to surge also allows larger waves to attack coastal structures causing beach erosion, overtopping, and flooding, and exerting buoyant uplift as well as wave forces on structures that are not normally subjected to such forces. Minimum water levels may also be of critical importance — for example, in the case of the stability of a waterfront bulkhead, which normally has a few feet of water at its base to help counteract soil pressures. Minimum water levels are of obvious operational importance in the siting of berthing and mooring structures.

The storm tide is the water depth that is added to the reference water level such as the mean high tide level (M.H.W.) or possibly the high water on spring tide (H.W.O.S.T.) to give the D.W.L., as shown in Fig. 5.6. The storm tide consists of several components, the most important of which are the wind-stress, barometric, and wave set-up, which are modified by coriolis effect, storm characteristics, local topography and currents, and possible resonant effects.

The total D.W.L. can be found by a vector addition of the various components as follows:

$$\text{D.W.L.} = d + \vec{A_s} + \vec{W_s} + \vec{P_s} + \vec{W_w} \qquad (5.1)$$

where d is nominal water reference depth, (i.e., M.L.W.); A_s is the astronomical tide at the time of the surge; W_s is the wind set-up, usually consisting of both an onshore (W_{sx}) and an alongshore (W_{sy}) component; P_s is the pressure or barometric set-up; and W_w is the wave

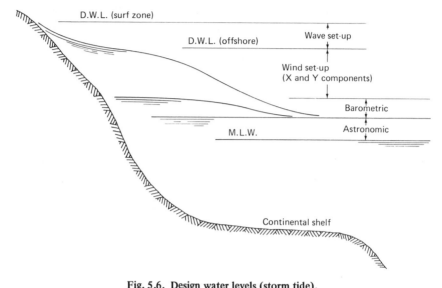

Fig. 5.6. Design water levels (storm tide).

set-up. Note that these terms are added vectorially and are nonlinear, as each term is dependent upon the vector sum of all the other terms, and the peak heights given by each term may occur out of phase. In estuaries and river mouths another term may be added to account for river runoff associated with heavy rains.

Wave set-up is only important along sloping coastlines and beaches where the energy of breaking waves allows for a local rise in water level. Wind stress is usually the most important single factor in piling up water against a coast. The above factors and some modifying effects are discussed individually in the following sections. As previously mentioned, storm surge is peculiar to shallow water sites and areas of broad gently sloping continental shelves such as found on the eastern and Gulf coasts of the United States; so; most of the examples given herein are related to these areas. Figure 5.7, after Harris,[13] shows hourly storm surge heights for various locations along the U. S. Atlantic coast during the passage of Hurricane Carol in 1954. As maximum water levels generally occur infrequently and on a random basis, the selection of design water level usually requires the use of statistics and probability theory and review of historical records where they exist.

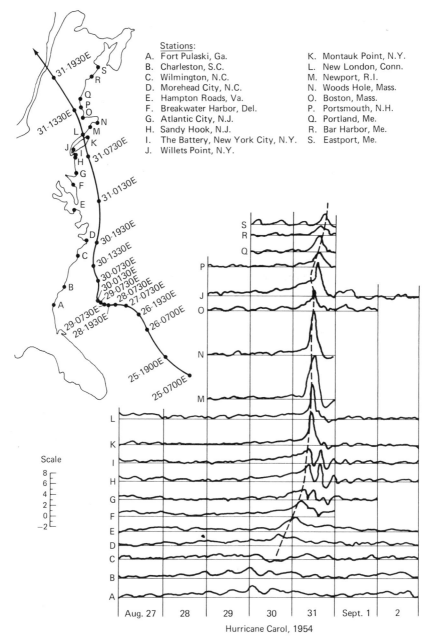

Stations:
A. Fort Pulaski, Ga.
B. Charleston, S.C.
C. Wilmington, N.C.
D. Morehead City, N.C.
E. Hampton Roads, Va.
F. Breakwater Harbor, Del.
G. Atlantic City, N.J.
H. Sandy Hook, N.J.
I. The Battery, New York City, N.Y.
J. Willets Point, N.Y.

K. Montauk Point, N.Y.
L. New London, Conn.
M. Newport, R.I.
N. Woods Hole, Mass.
O. Boston, Mass.
P. Portsmouth, N.H.
Q. Portland, Me.
R. Bar Harbor, Me.
S. Eastport, Me.

Hurricane Carol, 1954

Fig. 5.7. Hourly storm surge height (observed minus predicted sea level), Atlantic coast tide stations, August 27–September 2, 1954. (After Harris.[13])

Frequency and Probability

The return periods of storm surges along the coast are directly related to the return periods of severe storms. The storms producing the highest surges are generally tropical cyclones, known as hurricanes in the United States. However, in certain localities subject to frequent and severe extratropical cyclones, such as New England, maximum high water levels may be associated with such storms,[14] or other locally severe weather conditions. Although hurricanes generally have lower central pressures and stronger winds, in the mid-latitudes extratropical lows can provide a much greater fetch length and persist for a significantly longer period of time than the peak surge associated with hurricanes. Gofseyeff and Panuzio[15] studied and compared the extreme high tide frequencies versus return period for surges associated with both tropical and extratropical storms for New York Harbor. The longer the duration of the surge-producing forces, the greater the likelihood of the peak surge occurring at the time of high tide. DeYoung and Pfafflin[16] found that the probability of occurrence (P) of extreme high water nearly follows a log-normal distribution, and that the return period (R_T) is given by the relation:

$$R_T = \frac{1}{12 \ (1 - P_1) \ (1 - P_2)} \tag{5.2}$$

where $(1 - P_1)$ = probability of exceedance of the astronomical tide, and $(1 - P_2)$ = probability that the wind set-up will be exceeded. The above equation assumes that any excess water level above the astronomical tide height does not occur until the normal tide has risen to its full height.

Figure 5.8 illustrates the results of a detailed study[14] of extreme tide heights at Portland, Maine. Three plots of tide height versus return period are given: for the maximum observed tides, for the maximum observed surge heights combined with the astronomical high tides, and for the case of all surges combined with all tides not assuming simultaneous occurrence. These results seem to suggest that superposition of the maximum surge heights with astronomical high tide should give reasonably conservative water levels for most structure design lives where long periods of record are not available. Figure 5.9, also from reference 14, shows two marigrams for Portland,

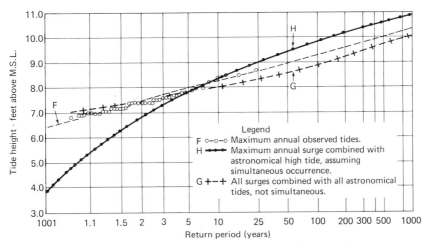

Fig. 5.8. Estimated probability of extreme high tide height at Portland, Maine. (Based on data for 1914-1959.) (After Peterson and Goodyear.[14])

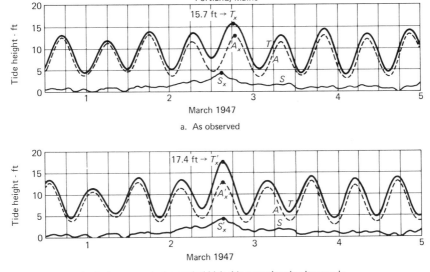

Legend

S — Storm surge - feet
A — Astronomical tide $\}$ Feet above gage zero
T — Total tide — $\big\}$ (−8.1 ft M.S.L.)

Fig. 5.9. Storm surge, astronomical tide and total tide, March 1-5, 1947 (a) as observed, (b) assuming peak surge and astronomical high tide occur simultaneously. (After Peterson and Goodyear.[14])

Maine, which illustrate the importance of the coincidence of peak surge with time of astronomical high tide. The upper marigram shows the total tide as actually observed during a winter storm and the lower marigram the height to which it would have risen if the peak surge had occurred simultaneously with astronomical high tide.

The U. S. Army SPM[17] lists maximum observed storm tide heights for many U. S. coastal locations over varying periods of record. Where historical records are unavailable, it would seem reasonable to superpose the peak surge height predicted from some probable severe storm condition for the area with the mean high tide or high water on the average spring tide. Note, however, that the actual observed astronomical tide will generally be lower and occur earlier than that predicted during a storm surge. An initial set-up of perhaps a foot or more is often observed,[18] prior to the arrival of a storm, which remains constant throughout the duration of the storm.

Extreme high tide frequencies for the U. S. East Coast associated with various return periods are shown in Fig. 5.10. These curves were presented by Ho[19] based on extensive studies conducted under the federal flood insurance program. This study considered the frequency and severity of both hurricanes and winter coastal storms using a

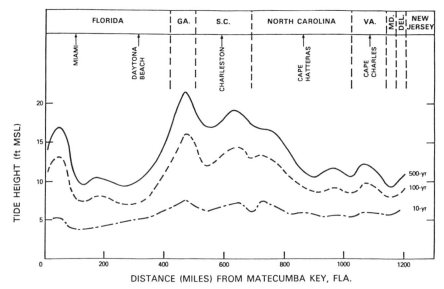

Fig. 5.10. Extreme tide frequency profiles along Atlantic Coast. (After Ho.[19])

computer simulation,[20] combining each surge with the astronomical tide in 84 different ways. For each of several locations the frequency of land-falling hurricanes, their central pressure and forward speed, and a shoaling factor were input. The shoaling factor accounts for the width and slope of the continental shelf, which controls certain dynamic enhancement of the wind-stress component of surge. This dynamic effect has been shown to be greatest for shelf widths of from 1/2 to 1/6 of the fetch length and for storm forward speeds that nearly coincide with the shallow water wave celerity.

Wind Set-up

The effect of wind in creating surface currents was discussed in Section 4.1. In deep water, the net transport of water at the surface is balanced by a return flow at lower levels. In shallow water, however, this return flow is impeded by bottom friction, and a pile-up of water against the coast will build until the pressure head thus developed is balanced against the resistance of flow caused by the bottom friction. Sibul and Johnson[21] studied the effects of bottom roughness on wind tides and confirmed that greater set-ups occur in shallow water and over rougher bottoms.

Wind-stress tides are not only caused by onshore winds, but wind currents moving parallel to a coastline may cause an additional pile-up of water due to deflection by the earth's rotation. This component of wind set-up is called the coriolis or bathystrophic set-up. Further, after a storm's passage, the coriolis tide component may persist if the longshore current continues as an inertial current (see Section 4.1). Wind-stress effects along the coast are illustrated in Fig. 5.11.

The general equation for the slope of the sea surface due to a surface wind stress is, neglecting higher-order terms;

$$\frac{dz}{dx} = \frac{\tau_s + \tau_B}{\gamma d} \tag{5.3}$$

where dz/dx is the sea surface slope, τ_s is wind shear stress, τ_B is bottom shear stress, γ is the unit weight of water, and d is mean water depth. These terms are clarified in Fig. 5.12 where it is shown that the difference in water level maintained by the wind shear and bot-

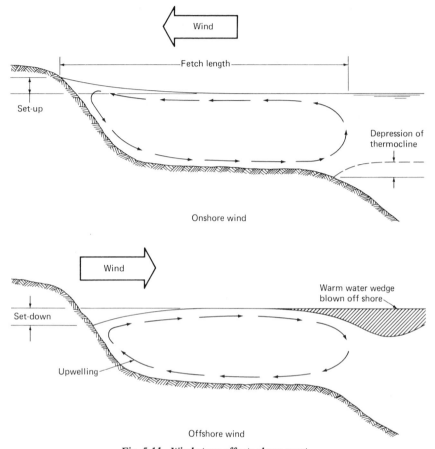

Fig. 5.11. Wind stress effects along coast.

tom friction along the direction of the fetch is balanced by an excess in hydrostatic pressure.

This equation can be further simplified for experimental reasons[22] by the introduction of a nondimensional parameter, λ, that relates the surface shear and bottom shear such that:

$$\frac{dz}{dx} = \lambda \frac{\tau_s}{\gamma d} \tag{5.4}$$

where λ has an overall average value of 1.27 with an extreme range from .7 to 1.8, and the value of τ_s can be approximated from the relation used by Sibul and Johnson:[21]

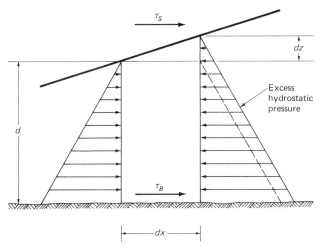

Fig. 5.12. Definition sketch for wind set-up.

$$\tau_s = 1.4 \times 10^{-6} \, V_{30}^{2.22} \qquad (5.5)$$

where τ_s is given in lb-ft^2 for the wind velocity as measured at 30 feet above the sea surface (V_{30}) given in ft/sec.

There are other semi-empirical formulas for estimating wind set-up, but an exact solution for both the two- and the three-dimensional case is very sophisticated and nearly impossible without the use of a high-speed digital computer. The engineer must rely upon the specialist oceanographer and meteorologist and upon historical records where they exist for this information. However, the engineer should understand the principles involved in order to assess the potential for wind tide to develop at a given site. The SPM[17] gives instructions for calculating the two-dimensional case, and presents curves for estimating the total storm surge along the U. S. eastern and Gulf coasts and gives examples for the Great Lakes for storms of various intensity. For a more detailed introduction to storm surge calculations, the reader is referred to the work of Bodine[23] and of Bretschneider.[24] The bathystrophic approach of Freeman et al.[25] is based upon numerical, finite difference techniques requiring the application of high-speed computers.

Calculation of storm tide also involves the selection of a storm wind field such as given by the standard project hurricane (SPH) or probable maximum hurricane (PMH), already introduced in Section

2.2. In addition, these references give information useful for developing project wind fields for storms of lesser intensity but perhaps higher frequency that may form the basis for design of structures of secondary importance or of short design lives, or of flexible type shore protection structures. As discussed in Section 2.2, the important parameters in storm surge prediction are central pressure index (CPI) or the reduction in barometric pressure, the radius of extreme winds or the distance from the storm center to the area of maximum wind speed, and the storm's forward speed and direction. These parameters in turn define the wind speed and fetch lengths required to calculate the wind stress set-up. In addition, the wind stress current velocity must be determined[26] so that the longshore component of current and hence the coriolis set-up can be calculated.

Pressure Set-up

A reduction in atmospheric pressure from the normal or ambient air pressure allows for a local rise in sea level. At the center of an SPH, this rise due to pressure drop alone (CPI) can be as high as 3 feet. As the specific gravity of mercury equals 13.6, a one-inch drop in the mercury of a barometer corresponds to a 13.6-inch rise in sea level. For a stationary storm system the pressure set up (P_s) in feet at the storm's center is given by:

$$P_s = 1.14\, \Delta P \qquad (5.6)$$

for ΔP given in inches of mercury.

The pressure reduction, ΔP_r, at any distance r from the storm's center can be found from:

$$\Delta P_r = \Delta P\, (1 - e^{-R/r}) \qquad (5.7)$$

where R is the radius to extreme winds for an SPH as defined in Section 2.2.

A moving storm system may cause dynamic enhancement of the pressure set-up when the storm forward speed approaches the shallow water wave speed $(C = \sqrt{gd}\,)$. Bretschneider[27] describes the response factor:

$$R = \frac{gd}{V_F^2 - gd} \qquad (5.8)$$

where V_F is the storm forward speed, from which it can be seen that when $V_F = \sqrt{gd}$, resonance occurs and the response factor goes to infinity. The fact that infinite amplication does not actually occur is due to damping effects, friction, varying water depths and storm speeds, and the short time element involved. However, there are locations (e.g., Narragansett Bay, Rhode Island) that are particularly prone to this type of resonance because of their relatively uniform water depths, which give wave speeds in the range of usual storm speeds and are roughly aligned with commonly followed storm tracks.

Wave Set-up

Waves breaking along a shoreline will cause an increase in the still water level (S.W.L.) that is separate from the run-up of water from the breaking waves and separate from any of the forms of set-up previously discussed. It is due primarily to the phenomenon of mass transport as wave energy is propagated toward the beach. This phenomenon exists primarily in the surf zone; so in general it applies only to the design of shore protection structures. Based upon the work of Saville[28] and the later work of Longuet-Higgins and Stewart[29] the Army Shore Protection Manual[17] gives the following formula for wave set-up:

$$W_w = .19 H_b \left[1 - 2.82 \sqrt{\frac{H_b}{gT^2}} \right] \qquad (5.9)$$

where H_b is the breaker height, and T is the incident wave period. For most design conditions this formula will give set-up values of around $.15 H_b$. However, more recent studies described by Hansen,[30] based upon field measurements on the island of Sylt in the North Sea, give maximum set-up values at the set-up line of 30% of the incident significant wave height and up to 50% of the significant breaker height. The maximum wave set-up was found to vary with the state of the tide, distance along the wave profile, type of breakers, and bottom slope. In general, the steeper the bottom slope, the steeper the set-up.

5.3   LONG-TERM EFFECTS

Long-term effects due to climatological and geological events occur at rates, generally years or thousands of years, that are seldom of concern to the structural designer. However, in certain localities climatological and geological processes may cause significant changes in relative sea level over relatively shorter periods of time. The Great Lakes are subject to seasonal and annual fluctuations of up to a foot or more associated with yearly rainfall patterns. Many coastal areas are subject to relatively sudden changes in relative sea level caused by subsidence of the land, due, for example, to: sediment consolidation, especially in deltaic areas; seismic activity, which may result in catastrophic ground motions; and the longer-term isostatic rebound of the earth's surface as it slowly responds to postglacial unloading. Anticipated rates of sea level change associated with the aforementioned effects are generally highly localized and can only be estimated by geological specialists.

Based on tide gage records from all over the world, it appears that the overall sea level is rising at a mean rate of approximately .04 inch per year.[31] The problem of determining whether sea level is rising or the land is subsiding complicates this picture considerably. References 32 and 33 discuss the overall eustatic rise in sea level on a more scientific basis. On a more practical basis of interest to the coastal and structural engineer, Hicks and Shofnos[34] have summarized annual sea level changes around the United States during the twentieth century. Marmer[35] has studied tide gage records along the U. S. Atlantic, Pacific, and Alaskan coasts for the first half of this century, which indicate that all locations exhibit seasonal as well as yearly variations on the order of a few to several tenths of a foot. The Atlantic and Pacific tide gages indicate a general overall rise in sea level with time, while along the southern Alaskan coast the trend is toward an overall fall in sea level. Estimates of the present long-term rates of sea level rise for the North Atlantic, Pacific, and Gulf of Mexico are .13 inch/year, .02 inch/year, and .08 inch/year, respectively.

Bruun[36] presented a discussion of shoreline erosion as related to sea level rise and found that the horizontal distance the shoreline retreats is approximately equal to the product of the sea level rise times the cotangent of the bottom slope.

5.4 BASIN OSCILLATIONS

General Description

As discussed in connection with the amplification of tides in bays and estuaries, enclosed and semi-enclosed basins of water have some fundamental period, and harmonic periods, at which the water in the basin will tend to oscillate when excited by some disturbing force. This can be readily demonstrated by tilting a rectangular pan of water and setting it back down and observing the back-and-forth sloshing motion. Lakes, harbors, bays, and so on, may be subject to such a sloshing motion when the water surface is initially displaced by some disturbing force, such as: atmospheric pressure fluctuations; ambient wave motion, especially that of irregular wave trains at the mouth of a semi-enclosed harbor; the tilting of the water surface by wind stress and its subsequent release; and, less commonly, strong currents and local seismic activity. The motion may be greatly amplified when the exciting force repeats itself at or near the fundamental period or one of the harmonic periods.

For a narrow rectangular basin, closed at both ends, as shown in Fig. 5.13, with length L and depth d, the transit time for one wavelength equal to twice the basin length is given by:

$$T = \frac{2L}{C}$$

where C is the wave celerity, which for most observed harbor oscillations can be demonstrated to be very nearly the shallow water wave speed, $C = \sqrt{gd}$. Therefore, the period of free oscillation of an enclosed, narrow rectangular basin is given by:

$$T_{nc} = \frac{2L}{n\sqrt{gd}} \tag{5.10}$$

where n is the number of nodes, which is equal to 1 for the fundamental period and takes on values of consecutive integers ($n = 1, 2,$

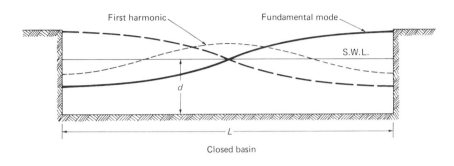

Closed basin

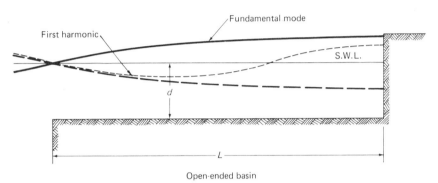

Open-ended basin

Fig. 5.13. Free oscillations of narrow rectangular basins.

3 . . . .) for the higher harmonic modes. The amplitudes tend to be lower for the higher harmonics (shorter periods), and in general the higher modes are discouraged by greater effective damping. In the case of an open-ended rectangular basin, referring again to Fig. 5.13, a node is effectively formed at the open end and the natural period becomes twice as long as that for a closed basin, or:

$$T_{no} = \frac{4L}{n \sqrt{gd}}$$ (5.11)

where n takes on values of only the odd-numbered integers ($n = 1$, 3, 5 . . . .). For an open rectangular basin that varies uniformly in depth from d at the opening to zero at the inshore end (i.e., is wedge shaped) the fundamental period becomes:

$$T \approx \frac{5.2L}{\sqrt{gd}}$$ (5.12)

The effect of friction is to cause a slight decrease in the natural period of the system. Basin oscillations are analogous to mechanical vibrating systems. When the period of excitation (T) is very much greater than the natural period (T_n) or $T_n/T \approx O$, the system response will have approximately the same magnitude and phase as the exciting force; when the exciting period is less than the natural period $(T < T_n)$, the amplitude of motion tends to be diminished; and when $T = T_n$, resonance occurs and the amplitude would theoretically go to infinity in a few cycles if damping were not present (see Fig. 5.14). In the case of a harbor being excited by a series of irregular wave trains at its mouth, it will selectively amplify those periods around its natural period and may also undergo forced oscillations at other dominant periods.

When the width of a harbor is relatively large with respect to its length, very complicated three-dimensional patterns of oscillation may

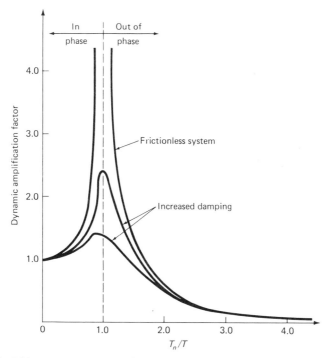

Fig. 5.14. Dynamic response of single degree of freedom spring-mass system.

be set up. Figure 5.15 illustrates the simple three-dimensional binodal case for a square enclosed basin for which the fundamental period is given by:

$$T = \frac{\sqrt{2}L}{\sqrt{gd}} \qquad (5.13)$$

Harbor entrance widths, geometries, variable depths, attached basins, and other features all affect the natural period of oscillation. For a general treatment of the problem of harbor surging and of more complex basin oscillations and its theoretical development, the reader is referred to Wiegel[37] and to Raichlen.[38]

In general, horizontal water velocities will be greatest at the node locations. It can be shown that for the simple two-dimensional case of the narrow rectangular basin, the maximum horizontal velocity, $V_{H\max}$, will be:

$$V_{H\max} = \frac{H}{2} \sqrt{\frac{g}{d}} \qquad (5.14)$$

where H is twice the amplitude (equal to the crest-to-trough distance). The velocities given by equation (5.14) are usually quite low, but the corresponding displacements may be quite large and when coupled with the long period can cause serious problems with moored ships located near the nodal points. The horizontal displacement about

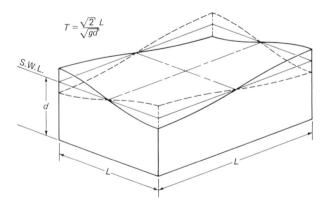

Fig. 5.15. Three-dimensional, binodal oscillation of square basin.

the node can be found by multiplying the average velocity by the half period, or:

$$X_H = \frac{HT_n}{2\pi} \sqrt{\frac{g}{d}} \qquad (5.15)$$

The simplified basic relations (equations 5.10 through 5.15), given herein for the idealized cases illustrated, should give the reader a feel for some of the basic governing factors and orders of magnitude of periods and motions. Observed harbor oscillations usually can be approximated by these relations. Understanding the basics of basin oscillations is important in determining the potential for structural loadings and operational difficulties at certain sites prone to such long wave motions. It must be borne in mind, however, that location of nodes, amplitudes, periods of motion, and so forth, for actual harbors can usually be ascertained only by field observations and sometimes by hydraulic modeling.[39]

The most common disturbing forces for lakes and nearly enclosed harbors are atmospheric pressure variations. Wilson[40] describes a relatively severe harbor seiche in Table Bay, Cape Town, South Africa, in response to a low pressure system moving rapidly eastward. Some of the most notable harbor oscillations, known as seiches or harbor surging, may occur on windless days when the sea surface is otherwise calm. The term seiche was originally used to describe the periodic change in water level observed in Lake Geneva, but it has come into general use to describe almost any form of long-period harbor oscillation. The term harbor surge or surging is also often used to describe the same phenomenon although Wiegel[37] suggests the use of "surge" for a once-only motion. Wilson[41] has also studied the large-scale resonant response of the continental shelf waters off San Pedro Bay, California to tsunami excitation.

Another common cause of seiche is waves, produced by: resonance with recurring periods in irregular wave trains, to reflected waves that become "trapped" inside a harbor with steep relatively impermeable sides, and motions set up by the wakes of large vessels entering or leaving a harbor. Theoretical investigation of the above is best left to the experts, but the designer or planner of harbor structures must be aware of when to consult others and when to conduct field studies.

Depending upon local climatic and geographic conditions, seiching may be only a seasonal or occasional occurrence in some locations, whereas in others it is a more-or-less regular feature.

Basin oscillations are essentially a long wave phenomenon having periods from perhaps 40 seconds to several hours. The wave spectrum is apparently fairly flat between the periods of ordinary wind waves (4–20 seconds) and observed long wave effects. Periods falling in between these ranges are a more complicated combination of the effects just discussed and those covered in Chapter 3. Most harbors with noticeable seiche have periods in the range of several minutes to 1 or 2 hours. Selected references regarding harbor surge problems of particular engineering interest include references 42 through 45.

One further special case of harbor oscillation is worthy of mention. It affects basins open to the sea through an inlet and is analogous to the Helmholtz resonator in acoustics whereby the water surface in the basin uniformly rises and falls while the water in the inlet channel moves back and forth. This mode of oscillation has periods typically greater than the fundamental modes previously discussed and has been shown to be significant for several harbors on the Great Lakes.[46]

Considerations in Structural Design

It has been mentioned that the large horizontal displacements of water at the nodal points and long periods that are unfortunately within the range of natural periods for large moored vessels may cause serious problems with moored ships. This is of obvious importance in the location of berthing and mooring facilities. Also the large movements can result in damaging shock loads to mooring structures and their appurtenances. The terminal forces developed in the mooring lines for a given amplitude of motion depend upon the elasticity of the lines, the initial tension, the mooring line geometry, and vessel displacement.

The problems of moored vessels and mooring line loads were discussed briefly in Section 3.6, but in general they are beyond the scope of this text. Wiegel[37] provides a more detailed introduction to the subject and presents some of the more salient results of various investigations into mooring line forces. Reference 47 contains several excellent papers on the response of moored vessels to long period wave action.

The vertical amplitudes of seiche motion are usually small, on the order of a foot or less, and are not usually significant in determining design water levels.

In summary, harbor oscillations may affect the siting (avoidance of node locations) and horizontal loads in structures that have floating bodies moored alongside. Floating breakwaters should also be investigated in this respect because they are normally designed to have periods much longer than the ordinary wind waves that they are designed to protect against, thereby making them susceptible to longer period waves. Finally, the placement of a new breakwater, fixed or floating, and/or other harbor or shore protection structures may create a seiche problem that did not exist before!

5.5 TSUNAMIS

General Description

Tsunamis are impulsively generated, dispersive waves of relatively long period and low amplitude. They are typically generated by sudden large-scale sea floor movements usually associated with severe, shallow-focus earthquakes. Usually an earthquake of at least 6.5 to 7.5 Richter magnitude and with focal depths of less than 30 or 40 miles is required to initiate such sea floor movements. Tsunamis may also be generated by underwater landslides, volcanoes, or explosions. Tsunamis can be highly destructive waves, especially at certain locations prone to tsunami run-up. Although they are almost undetectable at sea, because of their long wavelengths, with periods of a few minutes to an hour or more, and heights of only 1 or 2 feet or less, when they approach shallow water, shoaling, refraction, and possible resonant effects can cause run-ups of tens of feet to perhaps 100 feet or more, depending on the tsunami characteristics and the local topography. Tsunamis often manifest themselves in a series of highly periodic surges that may continue over a period of several hours. Since the greatest seismic activity observed rims the North Pacific Ocean, the most destructive tsunamis have been recorded here. Susceptible areas of engineering importance are typically low-lying coastal regions suitable for shoreline development. Bascom[11] recounts some historical records of tsunami damage, and there have been several engineering investigations of tsunamis in this century, notably references 48–50.

Because of the very great wavelengths (100 miles or more) tsunamis travel everywhere at sea at very nearly the shallow water wave speed: $C = \sqrt{gd}$. Therefore, when the location of the epicenter of the earthquake is known, the travel time of a tsunami between points can be fairly accurately predicted using nautical charts by summing the incremental travel times along the wave fronts, or:

$$t = \sum \frac{\triangle S}{\sqrt{gd_s}} \qquad (5.16)$$

where t is the travel time, and $\triangle S$ is the spacing of wave orthogonals or the incremental distance over which d_s, the depth, is relatively constant. If we consider the average depth of the Pacific Ocean to be about 12,000 feet, then the tsunami wave will travel at an average speed of nearly 400 knots. Charts of tsunami travel times from all possible sources for locations frequently attacked by tsunamis have been developed.[51, 52]

Damage due to tsunamis is from large hydrostatic and hydrodynamic forces and impact of waterborne objects, overtopping, and consequent flooding, and from erosion caused by the high water velocities. Van Dorn[53] reports that representative bore heights of 10 to 20 feet and flow velocities of 30 to 50 feet per second have been observed, and also provides an introduction to tsunami engineering. Wiegel[54] has reviewed the frequency and probability of occurrence of tsunamis for some U. S. locations based upon historical data, and also provides an excellent introductory discussion of tsunamis.

Engineering Aspects

The characteristics of tsunamis of primary interest to structural engineers are the nearshore properties (i.e., run-up heights, surge or bore velocities, and periodicity). Tsunamis may also excite some harmonic frequency of certain harbors,[41] such as discussed in Section 5.4, which may give rise to resonant oscillations lasting a day or more after the initial shock. Tsunamis have very low steepness and hence high relative run-up. Kaplan[55] conducted laboratory experiments aimed at predicting tsunami run-up and developed empirical relations for 1:30 and 1:60 plane slopes. He plotted his results in terms of relative run-up (R/H) and relative steepness (H/L) where

H and L are the incident wave height and length at the toe of the slope and R is the vertical run-up height above the still water level. Kaplan presented the following empirical relations:

For 1 on 30 slopes—

$$R/H = .381 \left(\frac{H}{L}\right)^{-.316} \qquad (5.17)$$

For 1 on 60 slopes—

$$R/H = .206 \left(\frac{H}{L}\right)^{-.315} \qquad (5.18)$$

Observed run-up heights may vary considerably from the ideal conditions studies by Kaplan. Bretschneider and Wybro[56] have done further studies of tsunami run-up introducing Manning's surface roughness coefficient in their inundation prediction equations and presenting curves of surge height versus distance inland for a given value of Manning's n, and for Froude numbers (F_N) equal to 1.0 and 2.0. When the surge velocity exceeds the long wave speed, $F_N > 1.0$, a bore may be formed which charges inland as a virtual wall of water. The possible formation of a bore is another important aspect of potential tsunami damage. The velocity of a surge can be estimated, according to Wilson and Tørum,[48] from the following equation:

$$V_s = K \sqrt{gd} \qquad (5.19)$$

where d is the depth at the front of the surge and k is a coefficient ranging from about 0.7 for high bed resistance to 2.0 for a frictionless bed.

Cross[57] has reviewed earlier works in the area of tsunami surge forces and compares velocities estimated from "postmortem" studies to those predicted by other investigators. Cross conducted laboratory experiments in order to evaluate tsunami surge forces assuming bore formation and that the shape of the bore front is known, and presented a semi-empirical formula for the maximum pressures. There are as yet no wholly satisfactory means of predicting forces and peak pressures on structures because of many interacting phenomena and the probabilistic nature of tsunamis. More work also remains to be done in the determination of when a bore will be formed, as the most destructive forces are developed under such conditions. From analysis of structural damage at Hilo, Hawaii, caused by the 1960 tsunami, Matlock et al.[50] determined that dynamic pressures of 400 to 1800 pounds per square foot occurred.

Tsunami hazard has historically been relegated to a few sites and has perhaps been neglected at others that exposure and topography make susceptible. It is, therefore, recommended here that the potential for tsunami damage be evaluated at virgin sites where important projects are planned. This would apply in particular to coastal or offshore nuclear plants where the potential for damage is great. Pararas-Carayannis[58] has recommended the equivalent of a standard project hurricane or maximum probable hurricane, which he calls the maximum probable tsunami (MPT), that would be based upon the largest possible seismic event that could give rise to such a tsunami for a given area.

REFERENCES

1. Doodson, A. T. "The Harmonic Development of the Tide Generating Potential," *Proc. Royal Soc.* A-100, 1922.
2. *Tide Tables*, 4 vols. covering most world locations, U. S. Dept. of Comm., NOAA, NOS, published annually.
3. Schureman, P. *Manual of Harmonic Analysis and Prediction of Tides*, U. S. Coast and Geod. Survey, S. P. #98, 1958.
4. Defant, A. *Ebb and Flow*, Ann Arbor Press, Ann Arbor, Michigan, 1958.
5. *Oceanographic Atlas of the North Atlantic Ocean*, Section 1, Tides and Currents, USNOO, H.O. #700, 1965.
6. Rude, G. T. "Tides and Their Engineering Aspects," *Trans ASCE*, Vol. 92, No. 1668, 1928.
7. Bowditch, N. (original work). *The American Practical Navigator*, Vol. 1, USNOO, H.O. #9, 1977.
8. Marmer, H. A. *Tidal Datum Planes*, USCGS, S.P. #135, 1951.
9. Disney, L. P. "Tide Heights Along the Coasts of the United States," *Proc. ASCE*, 81, separate no. 660, April 1958.
10. Harris, D. L., and Lindsay, C. V. *An Index of Tide Gages and Tide Gage Records for the Atlantic and Gulf Coasts of the United States*, U. S. Dept. of Comm., National Hurricane Research Project, NHRP, #7, 1957.

11. Bascom, W. *Waves and Beaches*, Doubleday, New York, 1964.
12. Perdikis, H. S. "Hurricane Flood Protection in the U. S.," *Proc. ASCE*, WW-1, Vol. 93, 1967.
13. Harris, D. L. *Some Problems Involved in the Study of Storm Surges*, U. S. Dept. of Comm., NHRP, No. 4, 1956.
14. Peterson, K. R., and Goodyear, H. V. *Criteria for a Standard Project Northeaster for New England North of Cape Cod*, U. S. Dept. of Comm., NHRP, No. 68, March 1964.
15. Gofseyeff, S., and Panuzio, F. L. "Hurricane Studies of New York Harbor," *Proc. ASCE*, WW-1, Vol. 88, Feb. 1962.
16. DeYoung, R. K., and Pfafflin, J. R. "Recurrence Intervals of Abnormally High Tides by Superposition of Storm Surges over Astronomical Tides," *Proc. ASCE*, Spec. Conf., Civil Eng'g. in the Oceans, III, 1975.
17. U. S. Army. *Shore Protection Manual*, Vol. 1, USACE, CERC, 1977.
18. Marinos, G., and Woodward, J. W. "Estimation of Hurricane Surge Hydrographs," *Proc. ASCE*, WW-2, Vol. 94, May 1968.
19. Ho, F. R. "Hurricane Tide Frequencies Along the Atlantic Coast," *Proc. ASCE*, 15th Conf. on Coastal Eng'g., Honolulu, 1976.
20. Barrientos, C. S. and Jelesnianski, C. P. "SPLASH – a Model for Forecasting Tropical Storm Surges," *Proc. ASCE*, 15th Conf. on Coastal Eng'g., Honolulu, 1976.
21. Sibul, O. J., and Johnson, J. W. "Laboratory Studies of Wind Tides in Shallow Water," *Proc. ASCE*, WW-1, Vol. 83, April 1957.
22. Keulegan, G. H. "Wind Tides in Small Closed Channels," *National Bureau of Standards, Journal of Research*, Vol. 46, #5, May 1951.
23. Bodine, B. R. *Storm Surge on the Open Coast: Fundamentals and Simplified Prediction*, USACE, CERC, TM-35, May 1971.
24. Bretschneider, C. L. "Engineering Aspects of Hurricane Surge," in *Estuary and Coastline Hydrodynamics*, edited by A. Ippen, McGraw-Hill, New York, 1966.
25. Freeman, J. C., Baer, L., and Jung, C. H. "The Bathystrophic Storm Tide," *Journal of Marine Research*, Vol. 16, 1957.
26. Bretschneider, C. L. "Storm Surges," Vol. 4, *Advances in Hydroscience*, Academic Press, New York, 1967.
27. Ibid. "How to Calculate Storm Surges Over the Continental Shelf," *Ocean Industry Magazine*, July 1967.
28. Saville, T. "Experimental Determination of Wave Set-Up," U. S. Dept. of Comm., NHRP, #50, 1961.
29. Longuet-Higgins, M. S., and Stewart, R. W. "A Note on Wave Set-Up," *Journal of Marine Research*, Vol. 21, 1963.
30. Hansen, J. "Wave Set-Up and Design Water Level," *Proc. ASCE*, WW-2, Vol. 104, May 1978.
31. Kuenen, Ph. H. *Marine Geology*, John Wiley & Sons, New York, 1950.
32. Fairbridge, R. W. "Eustatic Changes in Sea Level," *Physics and Chemistry of the Earth*, Vol. 4, Pergamon Press, New York, 1961.

33. Rossiter, J. R. "Long Term Variations in Sea Level," *The Sea,* Vol. I, Wiley Interscience, New York, 1962.
34. Hicks, S. D., and Shofnos, W. "Yearly Sea Level Variations for the United States," *Journal Hydraulics Div., ASCE,* 1965.
35. Marmer, H. A. "Changes in Sea Level from Tide Gage Measurements," *Proc.* 2nd Conf. on Coastal Eng'g., Council on Wave Research, Berkeley, 1952.
36. Bruun, P. "Sea Level Rise as a Cause of Shore Erosion," *Proc. ASCE,* WW-1, Vol. 88, Feb. 1962.
37. Wiegel, R. L. *Oceanographical Engineering,* Prentice-Hall, Englewood Cliffs, N.J., 1964.
38. Raichlen, F. "Harbor Resonance," in *Estuary and Coastline Hydrodynamics,* McGraw-Hill, New York, 1966.
39. Carr, J. H. "Long Period Waves or Surges in Harbors," *Trans. ASCE,* Vol. 118, No. 2556, 1953.
40. Wilson, B. W. "Generation of Long Period Seiches in Table Bay, Cape Town, by Barometric Oscillations," *Trans. Am. Geophysical Union,* Vol. 35, No. 5, Oct. 1954.
41. Ibid. "Tsunami-Responses of San Pedro Bay and Shelf, Calif.," *Proc. ASCE,* WW-2, Vol. 97, May 1971.
42. Ibid. "Origin and Effects of Long Period Waves in Ports," *Proc.* XIXth Int. Nav. Cong., London, Sept. 1957.
43. Ibid. "Seiches," Vol. 8, *Advances in Hydroscience,* Academic Press, New York, 1972.
44. Apte, A. S., and Marcou, C. "Seiches in Ports," *Proc.* 5th Conf. on Coastal Eng'g., Council on Wave Research, Berkeley, 1955.
45. McNown, J. S., and Donel, P. "Seiche in Harbours," *Dock and Harbour Authority,* 33, 384, Oct. 1952.
46. Sorenson, R. M., and Seelig, W. N. "Hydraulics of Great Lakes Inlet-Harbor Systems," *Proc. ASCE,* 15th Conf. on Coastal Eng'g., Honolulu, 1976.
47. "Analytical Treatment of Problems of Berthing & Mooring Ships," *Proc.* of NATO Advanced Study Institute, ASCE, 1965.
48. Wilson, B. W., and Tφrum , A. "The Tsunami of the Alaskan Earthquake, 1964; Engineering Evaluation," USACE, CERC, TM-25, 1968.
49. Magoon, O. T. "Structural Damage by Tsunamis," *Proc. ASCE,* Santa Barbara, Coastal Eng'g., Specialty Conf., 1965.
50. Matlock, H., Reese, L. C., and Matlock, R. B. "Analysis of Structural Damage from the 1960 Tsunami at Hilo, Hawaii," Report to Defense Atomic Support Agency, University of Texas, 1961.
51. Zetler, B. D. "Travel Time of Seismic Sea Waves to Honolulu," *Pacific Science,* Vol. 1, 1947.
52. Gilmour, A. E. "Tsunami Warning Charts," *New Zealand Journal of Geology and Geophysics,* Vol. 4, 1961.
53. Van Dorn, W. G. "Tsunami Engineering," *Topics in Ocean Engineering,* Vol. 3, edited by C. L. Bretschneider, Gulf Publishing Co., Houston, 1976.

54. Wiegel, R. L. "Tsunamis," *Earthquake Engineering*, Prentice-Hall, Englewood Cliffs, N.J., 1970.
55. Kaplan, K. "Generalized Laboratory Study of Tsunami Run-Up," USACE, CERC, TM-60, 1955.
56. Bretschneider, C. L., and Wybro, P. G. "Tsunami Inundation Prediction," *Proc. ASCE*, 15th Conf. on Coastal Eng'g., Honolulu, 1976.
57. Cross, R. H. "Tsunami Surge Forces," *Proc. ASCE*, WW-4, Vol. 93, Nov. 1967.
58. Pararas-Carayannis, G. "Tsunami Hazard and Design of Coastal Structures," *Proc. ASCE*, 15th Conf. on Coastal Eng'g., Honolulu, 1976.

6
Effects of Ice

In many areas of the world, including the Arctic, sub-Arctic and even Temperate Zones, marine structures must be designed to withstand the effects of ice. The type of forces exerted upon a structure will depend upon the nature of the ice, (thickness, concentration, mechanical properties, etc.) and the relative location of the structure (local climatology, currents and tide range, proximity of land, etc.). The design of a fixed offshore platform exposed to thick pack ice and intense pressure fields will likely be governed by ice loads, whereas in more temperate climates, light pile structures such as found in small craft harbors and marinas may be exposed to damaging ice effects relatively infrequently; so the probability of encounter of icing conditions must be weighed against the economics of potential damage. The nature and magnitude of ice forces will vary greatly with structure type and ice conditions (e.g., the tremendous lateral thrust of pack ice driven against an offshore structure, the impact of drift ice, or the vertical uplift, "jacking," of marina piles). Thus, ice loads exhibit high variability with climatic conditions, local physiography, properties of ice and structure type, configuration, and location.

This chapter will first describe the types and distribution of ice and the factors that contribute to ice growth and persistance. The mechanical properties and factors that affect the strength of ice are discussed. Finally, the types and magnitudes of forces exerted by ice on structures will be reviewed. Ice forces may be direct, such as thrusting and impact, or indirect, such as increased wind and gravity loads due to ice accretion. Other effects of ice such as abrasion and freeze–thaw damage are briefly discussed.

6.1 GENERAL DESCRIPTION AND DISTRIBUTION OF ICE

A significant percent of the earth's surface area remains perpetually frozen as the polar ice caps. Between the poles and higher latitiudes (say, approximately 40°) the ice found may vary from year-round ice of varying seasonal thickness to ice of minimal or moderate thickness, present only for 1 or 2 months of the year or perhaps only during infrequent cold winters.

Here we are concerned with sea ice and its distribution. Because ice in the sea presents obvious hazards to navigation, much of what is known concerning the description and distribution of sea ice is found in navigation publications such as those published by the U. S. Naval Oceanographic Office (USNOO). Reference 1 is one such publication, which provides an excellent introductory treatment to the subject of ice in the sea and describes means of observing, reporting, and obtaining ice data. The World Meteorological Organization (WMO) defines internationally accepted terms for the description of sea ice.[2] A few such terms are as follows:

New ice is recently formed ice composed of crystals weakly frozen together that have a definite form only while afloat; it is also called frazil ice, grease ice, slush, or shuga. Nilas is a thin elastic crust of ice that bends under wave action. Ice rind is similar to nilas but is more shiny and brittle. First-year ice accrues after not more than one year's growth and is typically less than 6 feet thick. Thin first-year ice is called white ice. Young ice is ice in transition between nilas and first-year ice and includes gray and gray-white ice. Old ice is ice that has survived at least one summer's melt. Brash ice is composed of accumulations of ice fragments up to 6 feet across. A bergy bit is a large piece of floating ice cleaved from a glacier showing less than 16 feet above the surface. A growler is smaller than a bergy bit and appears green or black in color. Fast ice, or land-fast ice, of particular importance in this text, remains fast against the shore to which it is attached. A floe is a relatively flat piece of sea ice 60 feet or more across. Floes may form from the rafting of ice cakes, which in turn may form from rafting pancake ice. Pancakes are individual circular pieces up to 10 feet in diameter and 4 inches thick. An iceberg is a massive irregularly shaped piece of ice broken from a glacier. Pack ice is any broad area of ice other

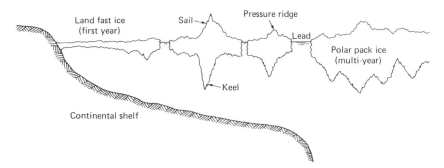

Fig. 6.1. Arctic ice—schematic cross section.

than fast ice. The concentration, or areal density, of pack ice is given in Oktas, or eighths of total coverage. Concentrations thus range from that of open water, which is considered as less than 1/8 coverage, to consolidated pack ice, which is 8/8 coverage. Lead is any navigable passage through sea ice, while polynya is any nonlinear opening enclosed in ice. Rafting refers to a pressure process whereby one piece of ice overrides another. Hummocks are hillocks of broken ice forced upward by pressure, which are called ridges when the upward-thrusted ice forms a line or wall. Rotten ice is honeycombed and in an advanced state of disintegration. A keel is a fin-like projection of ice projecting downward from the bottom of a floe or pack ice, while a sail is a similar projection upward from the surface of the floe. Some of these features are illustrated in Fig. 6.1, which represents a typical profile of ice conditions that may be found in the Arctic seas during the spring break-up.

Generalized charts showing the seasonal extent of ice in the polar seas and North Atlantic Ocean can be found in references 3 and 4. Aside from the perpetually frozen Arctic ice pack, floating ice within the Arctic Circle may attain a thickness of 12 to 15 feet, melting somewhat in the summer months. Figure 6.2, from reference 1, shows the mean seasonal thickness of fast ice for two typical sheltered harbors in the Northern Hemisphere, at the latitudes indicated.

Impact from drifting ice can be as hazardous to structures as thick fast ice. Icebergs are found as far south as 40° north latitude off the Grand Banks of Newfoundland through late spring and early summer. Drift ice is driven by the wind or by currents, and its path is de-

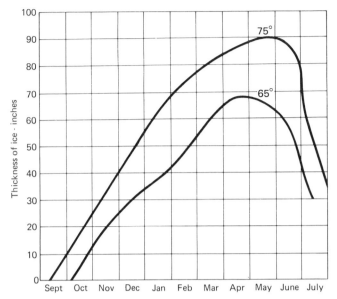

Fig. 6.2. Thickness of ice in two typical sheltered harbors in the northern hemisphere, at the latitudes indicated. (From U.S.N.O.O.[1] Reproduced by courtesy of the Defense Mapping Agency Hydrographic/Topographic Center.)

flected about 30° to the right of the wind direction (in the Northern Hemisphere) by coriolis force. The rate of drift is about 1% to 7% of the wind speed.

Table 6.1 gives the approximate rate of drift and drift angle from the wind direction for small and medium bergs as a percentage of wind speed. Detailed instructions for calculating the drift of icebergs as well as for short-term forcasting of ice conditions can be found in reference 5.

Ice floats with most of its volume below the sea surface, depending upon the relative density of the ice and water. Height-to-draft ratios of icebergs vary with the shape of the berg. In general, highly irregular shapes with spires and valleys have low ratios, perhaps as low as 1:1, while the usual height-to-draft ratio for tabular bergs with near-vertical sides is between 1:5 and 1:7. The ratio naturally also varies with the relative density of the ice and seawater.

The freezing point of seawater is depressed by its salinity, as shown in Fig. 6.3. Seawater of average normal salinity, about 35

Table 6.1. Drift of icebergs.

Wind speed (knots)	Ice drift speed (percent of wind speed)		Drift angle (degrees)	
	Small berg	Med. berg	Small berg	Med. berg
10	3.6	2.2	12	69
20	3.8	3.1	14	55
30	4.1	3.4	17	36
40	4.4	3.5	19	33
50	4.5	3.6	23	32
60	4.9	3.7	24	31

From USNOO.[1] Reproduced by Courtesy of the Defense Mapping Agency Hydrographic/Topographic Center.

parts per thousand (‰), freezes at about 28.6F°. Cooling water contracts to a maximum density at some temperature, beyond which it begins to expand again. The temperature of maximum density versus salinity is also indicated in Fig. 6.3. The temperature of maximum density is important in the freezing process because it places a limit on convective circulation. As surface water is cooled and contracts, it begins to sink, allowing water from lower depths to rise and be cooled. In shallow lakes and inland waters, the temperature of the entire water column must be lowered to about 39.4°F, the temperature of maximum density, before the surface will freeze, because of vertical convection currents that bring less dense water up from the bottom. Note also in Fig. 6.3 that the temperature of maximum density and the freezing point are equal at a salinity of approximately 24.7‰ with a corresponding temperature of about 29.6°F. This implies that for water of salinity greater than the critical value of 24.7‰ a convective circulation carrying warmer bottom water to the surface would continue until the entire water column was at the freezing point. This is not the case, however, because in polar seas the vertical distribution of salinity limits convective circulation and allows the surface to freeze. In very shallow seawater ice crystals sometimes will begin to form at the bottom or at mid-depth and eventually attach themselves to the bottom, a phenomenon known as anchor ice.

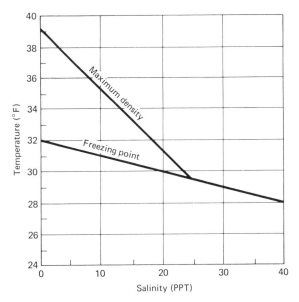

Fig. 6.3. Temperature of maximum density and freezing point of seawater. (Adapted after U.S.N.O.O.[1] by courtesy of the Defense Mapping Agency Hydrographic/Topographic Center.)

The Russian scientist N. N. Zubov[6] formulated an equation for the growth of ice in relation to air temperature based upon observations along the Russian Arctic coast. If sufficient temperature data are available for a given area, the thickness of ice to be expected can be estimated from Fig. 6.4, which is based upon Zubov's work. In this figure, a frost degree-day is considered to be the number of degrees below $32°F$ averaged for the day times the number of days over which the temperature was averaged. For example, if a mean temperature of $22°F$ was maintained over a period of 5 days, we would have accumulated $(32 - 22) \times 5$ days = 50 frost days for the period. This figure is convenient for estimating hypothetical ice thicknesses, but for structural design purposes it is always best to rely upon historical records where they exist or upon in situ measurements taken over as long a period of time as is possible and perhaps extrapolated on a statistical return-period basis.

For offshore structures situated in areas where ice hazards may exist, statistical ice data to be compiled for the specific location

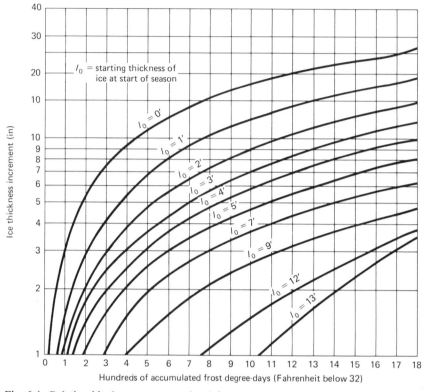

Fig. 6.4. Relationship between accumulated frost degree-days and ice growth for varying initial ice thicknesses (small degree-days accumulations). (After Zubov.[6])

should include information on the concentration and distribution, type of ice, mechanical properties, thickness, drift speed and direction, tide range, air temperature range, and probability of encountering icebergs.[7]

6.2 MECHANICAL PROPERTIES OF ICE

The physical and mechanical properties of sea ice are highly variable. They depend upon the temperature, the salinity, the rate of freezing, and, in the case of compressive strength, the rate of loading, the age of the ice, impurities, and other factors. Pounder[8] gives a thorough treatment to the physics of ice. It is particularly important to note that ice is, in general, an anisotropic material (i.e., it does not exhibit the same properties in all directions), although some fine-grained forms of ice are nearly isotropic and homogeneous. Sea ice consists of a large number of thin parallel layers, something analogous to a deck of cards, with inclusions of brine, air, and other impurities. These inclusions act as stress concentrators. The layers, or sheets, of ice crystal lie in a plane perpendicular to the weak axis of the ice. The so-called C-axis is the optical axis of the hexagonal ice crystals. Ice type and grain size are somewhat determined by the orientation of the C-axis.

Ice strength is dependent upon the volume of trapped brine. As the seawater begins to freeze, certain salts composed of the ions normally found in seawater begin to precipitate out. Therefore, normal salt water ice has a salinity of about 10‰ , whereas ice that has frozen very rapidly retains more salts and may have a salinity of 25‰. Such ice normally does not acquire significant strength until a temperature of +16°F or lower is reached. At +17.6°F, sodium sulfate (Na_2SO_4) begins to leach out, and the ice begins to acquire some strength. As the temperature is lowered below approximately –9°F, sodium chloride (NaCl) begins to leach out, and the ice acquires even more strength. In general, the colder the ice, the greater the tensile strength, and in general the greater the brine content the weaker the ice. Seawater ice is weaker than freshwater ice at high temperatures but may be up to twice as strong as freshwater ice at very low temperatures. Ice greater than a year or so old may be entirely free of salt as the salts tend to leach out with time. The strength of sea ice (σ) can be expressed in terms of the brine content of the ice in accordance with the following generally accepted formula:

$$\sigma = \sigma_0 \left[1 - \sqrt{v/C} \right] \qquad (6.1)$$

where σ_0 is the basic ice strength at zero brine volume, v is the relative brine volume, and C is a constant related to the geometric properties of the brine channels within the ice and other variables.

Failure of ice under load usually starts at some imperfection, the mode of failure (i.e., tension, compression, flexure, buckling, etc.) being all-important. Ice is relatively weak in tension, for example, the property most measured and used as an index of ice strength is the compressive strength. Published test results of ice properties exhibit wide scatter primarily due to the noncalibration of testing parameters (i.e., the variability of ice properties with temperature, salinity, rate of application of load, etc.), and perhaps the size and shape of specimen and type of test apparatus.

Table 6.2 was presented by this writer,[9] and represents a summary of sea ice properties of engineering interest based upon a review of various literature and values suggested by other authors, notably references 10–14. It is emphasized, however, that these values should be used for preliminary estimates only and/or for guidance when no other field or site-specific experimental data are available.

The type and strength of ice formed at a given site and under given conditions can be highly variable. The strength of ice is sensitive to rate of application of load, and to contact area; in general

Table 6.2. Properties of sea ice, suggested design values for general engineering purposes.*

Specific gravity	.86 to .92 (average values)
Compressive strength	400 to 600 psi (up to 3000 psi for pure freshwater ice)
Tensile strength	100 to 200 psi
Shear strength	Few test results
Modulus of elasticity	1.4×10^6 psi
Modulus of rupture	200 psi
Poisson's ratio	.35
Coefficient of thermal expansion	.000028 (average between $-20°$ and $32°$ F)
Coefficient of friction	.15 metal to sea ice .10 metal to freshwater ice .01 "wet" ice
Adhesion	30 to 100 psi
Volumetric expansion	9% (on freezing) maximum pressure exerted: 30,000 psi

*Above values to be used in lieu of observations or experimental data. Note well that the properties of sea ice are highly variable with respect to temperature, salinity, and the rate of freezing. Ice is an *anisotropic* material!

From reference 9

the slower the rate of load application, the higher the relative strength. This fact may be significant when selecting a value of compressive strength for calculating impact loads versus static thrust from large, slow-moving ice sheets. Ice is a viscoelastic material, however, so sustained loading produces time-dependent deformations. The rupture strength of ice decreases with increasing sustained load.[15]

Ultimate crushing strength tests on sea ice usually fall in the range of approximately 200 to 1000 psi, although higher and lower values have been reported. Values as high as 3000 psi have been reported for pure freshwater ice. Values of 400 psi[14] and 600 psi[13] have been proposed for the design of pile structures and ice breakers, respectively. Tensile strength tests usually fall in the range of 100 to 200 psi,[10] the lower value being more common. Flexural tests usually give values of the modulus of rupture, averaging slightly less than 200 psi.[10] I have little information on the shear strength of ice, but for most engineering structures ice usually fails by crushing or flexure or some combined condition, so that shear strength is not of major consequence. Estrada and Ward[13] used a value of 290 psi for shear strength in their discussion of ice forces on ships and found that shear strength was not of consequence. This value is probably conservatively high because the few values I have noted are more on the order of the tensile-strength values given above. The modulus of elasticity may vary from 250,000 psi to over 1,400,000 psi,[14] the lower value generally applying to higher temperatures and greater applied load.

Values of the coefficient of friction of ice on metal have come from the literature on icebreaker design.[13, 16] Note that "wet" ice only develops one-tenth or less the friction force of dry ice. Values of the static coefficient of friction of cold, dry ice on steel can be as high as 0.30 to 0.50[17]. So a higher value than shown in Table 6.2 should probably be used for Arctic conditions.

Adhesion of ice varies with the material to which it is attached. Freiberger and Lacks[18] give a range of values for woods of between 45 and 80 psi and for metals between 85 and 120 psi. Glass and plastics give an even wider range. Adhesion varies with temperature as well as with materials and may be limited by the shear strength of the ice.

Other values given in Table 6.2 should be self-explanatory. In addition, ice exhibits other interesting properties such as regelation. If two pieces of ice are brought into contact, very little applied pressure will cause them to melt together upon release of the pressure. If there is no change in the orientation of ice of the same structure, then there will be no noticable loss of strength of the separated ice. Thus, cracked ice sheets have the ability to heal themselves.

6.3 ICE FORCES

Some of the more important ways in which ice exerts forces on structures are summarized in the composite diagram of Fig. 6.5. These modes of loading will be discussed in turn. In particular, horizontal thrust due to moving or expanding ice sheets and vertical uplift loads on pile structures will be given attention.

In general, the magnitude of ice forces depends upon the mode of load application and/or the relative movement of the ice with respect to the structure, the flexibility and configuration of the structure, the characteristics of the ice cover, and the mechanical properties of the ice. In consideration of the above, functional design (i.e., designing to avoid ice forces) should be an important design consideration, as will be apparent in the following discussion.

Water that has been trapped in pockets or crevices will expand on the order of 9% in volume during the solidification process. Above a temperature of –9°F a pressure of about 30,000 psi would cause melting of the ice; thus reference 19 suggests this pressure as an upper limit for the freezing of captive water.

Ice accretion due to rain and freezing spray can considerably increase the exposed area on which the wind acts. Ice accretion due to spray alone may be on the order of 1 to 2 feet or more in thickness. Ice is slightly less dense than water; a mean unit weight of about 56 pounds per cubic foot seems a reasonable value for estimates, and can considerably increase vertical dead loads.

Ice exerts horizontal thrusts due to the pressure of an ice sheet being held against the structure by wind or current action, and/or by thermal expansion. Pressure fields develop within fast ice packs

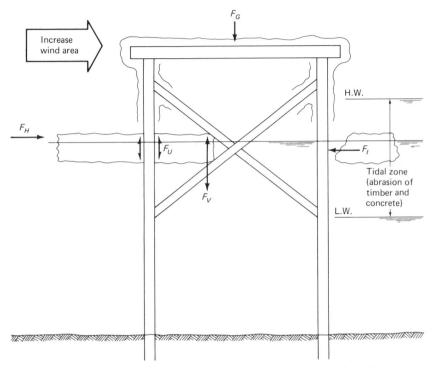

Fig. 6.5. Composite diagram of ice forces. Abbreviations: F_e = expansion of captive ice freezing in pockets (not shown); F_G = increase gravity load due to accumulation of ice and frozen spray; F_H = horizontal thrust due to pressure of ice sheet moved by wind or current stress or due to expansion of ice with rising temperature in enclosed area; F_I = impact of iceburgs and bits driven by wind or current (or vessels); F_U = uplift on piles due to adhesion of ice and rising water level; F_V = vertical force due to buoyancy of ice caught beneath X-bracing or batter piles and downward due to weight of trapped ice sheet. Note: F_H and F_I do not act simultaneously.

due to dynamic interaction between the ice-air and water interface. The resulting pressure gradients within the ice sheet may manifest themselves as a fluctuating force on the order of one or more cycles per second.   Therefore, transient vibrations and possibly dynamic amplification due to structure resonance could occur.

Icebergs driven by wind or currents may collide with a structure, thus imparting impact forces.   Ice frozen to piles and crossbracing may exert uplift forces on a rising tide, or ice trapped below structural members may exert direct buoyant uplift.

Peyton[15] presents an excellent introduction to the action of ice on marine structures. Gerritsen[20] gives an overview of ice problems and hydraulic structures. More recently, Brown[21] has summarized contemporary research into the mechanics of Arctic ice as related to ice forces on structures. Bercha and Stenning[22] demonstrate the application of Arctic ice mechanics to the design of monocone structures proposed for year-round petroleum production in 200 feet of water in the South Beaufort Sea. In general, additional information on the interaction of ice and sea structures can be found in the *Proceedings* of the International Conferences on Port and Ocean Engineering under Arctic conditions, which have been held bi-annually since 1971.

Horizontal Thrust and Impact of Ice

Ice may exert lateral forces on structures via various mechanisms. These forces depend upon many variables, such as the movement and extent of ice cover, the character of the cover and properties of the ice, and the flexibility of the structure itself. Large expanses of ice, such as rafted floes and dense-pack ice may thrust against a structure under the action of wind or current. Very high loads can be developed in this way. An example would be the design of oil platforms in Cook Inlet, Alaska, where the tide range is 35 feet or more and tidal currents can move in excess of 6 knots! Ice floes in this area may be over a mile in diameter and over 3 feet thick. Taking the compressive strength of ice for the site to be around 300 psi, the force exerted on a 200-inch-diameter column would be 2160 kips, which Peyton[15] points out may oscillate from zero to full load at a rate of one cycle per second.

Ice moving past a pile structure is gradually cut by the pile, leaving an open swath in the ice sheet. The force developed is dependent upon the rate of load application (i.e., as the rate of deformation increases the compressive strength decreases). Gerritsen[20] reports that Canadian design engineers have used a crushing strength of 280 tons/sq meter (approximately 400 psi) for the design of highway bridge piers, and adds that this value may be conservative based upon measurements taken over a 4-year period on two instrumented bridge piers, which yielded unit pressures from moving ice sheets in the

range of 70 to 113 tons/sq meter (100 to 162 psi). Current highway
design practice in the United States also calls for a design ice pressure
of up to 400 psi to be applied to bridge piers[23] depending upon
the type of ice conditions and configuration of the bridge pier.
Wortley[24] reaffirms that such values of ice crushing strength are
overly conservative and notes that the greatest pressures measured
in situ have been less than 200 psi. Based upon his extensive experience in the Great Lakes, Wortley[25] also notes that lateral forces
on vertical piles in small craft harbors are probably on the order of
or less than the mooring loads for which they have been designed.
A slight inclination of piles from the vertical helps reduce both
thrust and uplift forces by increasing the flexural stresses in the ice.

Higher compressive strengths generally correspond to lower rates
of load application. Therefore, very slowly moving ice may exert
the highest forces. For smaller floes and bergs moving at measurable velocities, the force exerted could be calculated by energy
methods if a mass, velocity, and coefficient of restitution between
ice and structure were assumed. Such information would be difficult to obtain with reasonable accuracy. Reference 19 suggests designing for impact loads on the order of 10 to 12 tons per square
foot of contact area for structures located in areas subject to drifting ice. Wortley,[25] however, based upon observations of pile-supported boat docks in protected harbors, suggests that the horizontal
forces from moving blocks of ice are less than the mooring forces
for which the piles were designed. Design loads are significantly
less than that given by the crushing strength of the ice, as the blocks
may impact against piles but not crush on them.

Dams[19, 26] and icebreakers[13, 16] have been designed to withstand
ice forces in the range of from 20 to 30 tons per lineal foot of ice
surface contact. For the case of a dam bordering an enclosed area,
the maximum thrust may be the result of spring thawing, as ice ex-
pands when warmed from lower temperatures still below freezing.
The thrust exerted by expanding ice sheets was studied by Rose,[27]
who presented curves of horizontal thrust versus ice thickness for
various rates of temperature rise assuming either zero or full lateral
restraint. The thrust is greater for the greater rates of rise in temperature and for the condition of complete restraint. Rose's work has
been used in the design of dams. However, more recent investiga-

tions, notably that of Drouin and Michel,[28] have somewhat super-
seded the work of Rose and others.

Drouin and Michel[28] conducted laboratory tests on both columnar and snow pack ice at various temperatures, and tabulated values
of horizontal thrust per unit length of ice sheets of various thicknesses, restrained in one direction for various rates of temperature
rise to 32°F. They found, for example, that for a 30-inch-thick ice
sheet with a surface temperature of +14°F warming up to 32°F in
10 hours, the horizontal thrust would be from 5 to 7 kips per foot
(Klf) for snow pack and columnar ice, respectively. For an initial
temperature of –22°F rising to 32°F in 20 hours, the corresponding
range of ice thrust becomes 20 to 30 Klf. In general, the average
corresponding contact stresses per unit area were less than 100 psi
with the maximum unit stresses occurring for the thinner ice, colder
initial temperatures, and slower rates of temperature rise. Stresses
are unevenly distributed within the ice, however, and are particularly
high in the upper 8 to 12 inches. Snow cover is a very effective
insulator, and a thickness of 6 inches or greater will prevent motion
of the ice.[24] Wortley[25] recommends a design value of approximately
10,000 lb per lineal foot, horizontal thrust on gravity type crib
structures in the Great Lakes.

In lieu of a more detailed analysis, reference 29 gives the following simplified formula for calculating the horizontal force (F_H) on
structures subject to ice hazard:

$$F_H = C_i f_{ci} A \qquad (6.2)$$

where A is the exposed area of the structure with regard to the thickness of the ice, f_{ci} is the compressive strength of the ice, usually
assumed to be between 200 and 500 psi,[29] and C_i is a coefficient
that accounts for shape, rate of load application, and other factors,
usually given a value between 0.3 and 0.7.

Icebreakers trapped in an ice field may be subject to strong pressure gradients within the ice. Because of the curved or angled shape
of the hull, a vertical force is also produced by the reaction of the ice
against the hull, the vertical force being dependent upon the coefficient of friction and the angle of the hull. An ice sheet acted upon
by a vertical load at its end is subject to a flexural stress like that on

a beam on an elastic foundation. Estrada and Ward[13] present an analysis of this condition and give an excellent review of the forces exerted by ice on ships. Nevel[30] solved the general problem of a semi-infinite plate on an elastic foundation loaded by either a distributed or a concentrated vertical force. Ice sheets may buckle because of the horizontal force and Hetenyi[31] and others have solved the problem of a beam on an elastic foundation acted upon by both a horizontal and a vertical force. Ice forces on large off-shore structures in Arctic regions are very much interrelated with the mechanics and behavior of the pack ice, such as ridging and pressure fields, and so on.[21, 22] The structural analysis and mechanics of failure of ice sheets are beyond the scope of this book. The reader is referred to the literature cited and to the various publications of the U. S. Army Corps of Engineers, Cold Regions Research and Engineering Laboratory (CRREL).

Changes in water levels cause tensile cracks to develop that fill with water which freezes and causes horizontal pressures as the water level rises. This force is greater in smaller enclosed areas such as tidal estuaries, and may approach the level of horizontal thrust associated with temperature rise.[14]

Figure 6.6 illustrates another manner in which destructive horizontal thrust forces may be developed in a small craft harbor in tidal water. Near the shoreline, ice along sloping banks is continually broken up by the rising and falling of the tide. The intact ice sheet seaward of this zone then can exert the full thrust developed

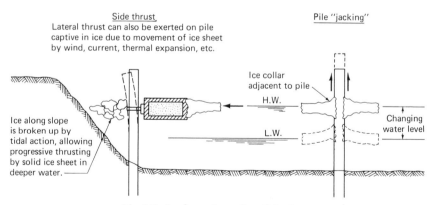

Fig. 6.6. Ice forces in small craft harbor.

by movement by wind or current, or expansion, etc., on structures situated in the transition zone. In this case the pile must resist the force transmitted along the entire length of the attached floats rather than only its own exposed area. It is desirable, therefore, to remove the marina floats from the water or moor them away from the fixed anchor piles in order to avoid damage to both piles and floats.

Icebreakers working in closed harbors may cause damage by forcing the ice sheet apart and thus setting up pressure gradients within the ice. Also, when the path of broken ice refreezes, it may cause some further movement of the ice sheet.

Small ice floes and broken pack ice may cause ice jams against a structure, resulting in a damming effect, which may create a differential head pressure. Ice jams are a particular hazard to navigation.

Aspects of functional design to minimize horizontal ice forces on structures include providing some degree of inclination in walls and piles to cause the ice to fail in bending before it can exert its full crushing strength. Crossbracing and vulnerable appurtenances should be avoided in the tidal zone because ice typically becomes jammed in the bracing and may shear off unsubstantial appurtenances. Figure 6.7 shows a functionally designed support pylon for an elevated walkway projecting into tidal water. The conical–elliptically shaped pylon should be effective in minimizing ice forces. The conical profile cleaves the ice on a falling tide, thereby reducing uplift, and the elliptical form tends to allow the ice to move by the pylon without exerting strong forces against its weak axis. Table 6.3 summarizes various effects of ice on structural design with suggestions for possible remedial measures.

Vertical Uplift Forces

Ice freezing around a pile subject to changing water levels will exert an upward force that is resisted by the structure's weight and the pile's pull-out resistance. Over a period of time, or over many tidal cycles, for example, piles without sufficient hold-down may be "jacked" upward by the continual rise and fall of the tide as illustrated in Fig. 6.6. Wortley[24] describes a method of calculating the

Fig. 6.7. Example of functionally designed small boat dock. Inset shows elliptical–conically shaped support pylon which results in ice forces significantly less than those for a traditional timber pile structure. *(Photo courtesy of Crandall Dry Dock Engineers, Inc.)*

254 The Marine Environment and Structural Design

Table 6.3. Summary — minimizing ice damage.

Effect of ice	Effect on structural design	Effect on functional design	Operational considerations
Thrust	Design for maximum expected thrust with regard to: ice thickness, temperature rise, degree of restraint, etc.	Avoid exposure to thrust if possible (i.e., moor floats with anchors vs. piles); use alternate structure, (i.e., cells vs. pier); use proper configuration.	Prevent freezing at structure, bubblers, heaters, etc.; if feasible, caution in use of ice breakers in confined areas.
Uplift (jacking)	Design for maximum expected uplift with regard to ice thickness, tide range, etc.	Minimize use and/or number of piles; design pile to resist uplift; use massive structure; minimize x-bracing; use proper shapes.	Use bubblers, moor floats in winter season; do not leave fixed to anchor piles.
Impact	Design for probable ice loads with regard to mass, velocity, and direction.	Avoid exposed sites.	Use camels or other protective devices.
Expansion	—	Avoid designing pockets and areas that can trap freezing water.	—
Gravity	Design for additional weight of ice with regard to expected accumulation.	Slope decks for drainage; use proper configuration.	Prevent ice from freezing on decks (i.e., heaters, salts?).
Buoyancy	Design for uplift as required.	Provide "air gap" above highest expected tide level; avoid x-bracing; design to prevent ice from getting under decks.	—
Accretion	Consider increased projected areas in wind and current force calculations; gravity loads as above.	Design with adequate drainage to prevent ice build-up.	Keep decks clear.
Abrasion	Consider effects of reduced section properties in design.	Use cladding, impervious materials; chamfer all sharp edges on concrete; avoid projections; protect exposed timber in tide zone.	—
Freeze–thaw (concrete)	Specify dense and durable concrete mix (i.e., low W/C ratio, type II cement); use air-entraining admixture.	Avoid immersion in tide zone (i.e., exposure to alternate wetting and drying).	—

minimum uplift force exerted by an ice sheet on a circular pile, based upon a "first crack" analysis, which assumes the formation of a circumferential crack at some distance from the center of the pile, termed the radius of load distribution, at which distance the slope of the deflected ice plate is zero. The ice then fails by cracking and not by slippage at the ice–pile surface. The radius of load distribution is greater than the pile diameter (approximately 6″ greater for steel piles and 3″ for timber piles) and is characterized in actuality by the formation of an ice collar or a zone of ice of thickness greater than the ice sheet adjacent to the pile. The method is a thin plate analysis, which may not be as accurate when the ice thickness is greater than the diameter of the pile plus the thickness of the ice collar. This method shows that the uplift force increases rapidly with the ice thickness but does not increase significantly with pile diameter. Therefore, the larger the pile, or structure, diameter, the smaller the relative uplift force. Assuming an ice flexural strength of 200 psi would give *minimum* uplift forces for a 12-inch-diameter pile on the order of 9, 30, and 70 kips (1 kip = 1000 lb) for ice thicknesses of 12, 24, and 36 inches respectively (slightly higher forces for steel and lower for timber piles). By contrast, the more exact uplift forces on a steel pile in 24-inch-thick ice would be approximately 33, 37, and 42 kips for pile diameters of 12, 24, and 36 inches, respectively. Corresponding adhesion values (applied to the pile perimeter × ice thickness) are typically less than 50 psi, but for very thick ice and smaller-diameter piles it is possible that the pile would slip before the ice failed. Although traditional methods of estimating ice uplift by applying an adhesion value such as indicated in Table 6.2 may have been conservative, Wortley cautions that the method described gives minimum values for design and maximum uplift loads could be up to several times greater, depending upon the ice strength and exact failure mechanism.

Where ice uplift is likely to be a problem, as in the design of light timber structures in tidal waters subject to seasonal ice thicknesses of a foot or more, it is desirable to maximize the pull-out resistance of the piles by driving them to as high a capacity as feasible, using modified pile tips, minimizing the number of piles, and maximizing the weight of structure carried by the piles as far as is compatible with other design requirements. It is also desirable to eliminate or minimize the use of crossbracing, as Peyton[15] notes that there have been cases of pier failures initiated by the sliding and prying action exerted against diagonal bracing and batter piles. Wortley[24] describes other methods of reducing ice uplift loads, including the use of low-friction surfaces and coatings. Piles that are successfully resisting uplift forces usually exhibit the formation of ice rubble,[25] or pile-up of ice debris clinging to the pile above the surface of the ice sheet that has accumulated from the successive destruction of ice collars.

The problem of ice uplift on walls and ice pressures on engineering structures in general is covered by Michel.[32] Modeling of important structures to determine ice forces and effects can be carried out in scale model ice basins.[33]

6.4 OTHER EFFECTS OF ICE

Ice causes abrasion of structural members in the tidal zone. Concrete is especially susceptible to abrasion and spalling, and timber piles are also sensitive to ice abrasion. Abrasion may increase the corrosion rate of steel structures by destroying protective coatings and films. Thus, tidal zone cladding should be considered in the design of susceptible structures. Ice may also carry highly abrasive materials within it and jam debris between structural elements, and ice may induce motions during thawing and possibly add to the impact forces of berthing ships. In perpetually frozen areas, consideration should be given to the fact that the presence of ice may preclude or greatly hamper maintenance and inspection operations. Freeze–thaw damage of concrete (see Section 7.5) is a serious problem related to icing. The continual freezing and thawing of water trapped in pockets of porous concrete as the water level varies eventually leads to progressive spalling and deterioration of the concrete.

Rubble stone structures and rip-rap protection are subject to ice "plucking" forces in the tidal zone. In general, such ice forces may be of the same order of magnitude as wave forces.[24]

REFERENCES

1. Bowditch, N. *The American Practical Navigator*, USNOO, H.O. #9, Vol. 1, 1977.
2. *Sea Ice Nomenclature*, WMO, No. 259, TP 145, Geneva, Switzerland.
3. U.S. Navy. *Oceanographic Atlas of the Polar Seas*, USNOO, H.O. #705, 1957.
4. Ibid. *Oceanographic Atlas of the North Atlantic*, Section 5, Sea Ice, USNOO, H.O. #700, 1968.
5. Wittman, W.I., and MacDowell, G. P. *Manual of Short Term Sea Ice Forecasting*, USNOO, SP-82, May 1964.
6. Zubov, N. N. "On the Maximum Thickness of Perennial Sea Ice," *Meteorologiia i Gidrologiia* 4, No. 4, 1938.
7. Det Norske Veritas. "Rules for the Design, Construction and Inspection of Fixed Offshore Structures," Oslo, 1974.
8. Pounder, E. R. *Physics of Ice*, Pergamon Press, Ltd., London, 1965.
9. Gaythwaite, J. W. "Structural Design Considerations in the Marine Environment," *Boston Soc. of Civil Eng'rs. Section, ASCE*, Vol. 65, No. 3, Oct. 1978.
10. Mantis, H. T., ed. *Review of the Properties of Snow and Ice*, USACE, SIPRE, Report No. 4, Wilmette, Ill., July 1951.
11. Peyton, H. R. "Sea Ice Strength," Univ. of Alaska, Arctic Environ. Eng'g. Lab., Geophysical Inst. Report, UAG R-182, Dec. 1966.
12. Weeks, W. and Assur, A "The Mechanical Properties of Sea Ice," U. S. Army Cold Regions Res. and Eng'g. Lab. (CREEL), Report II-C3, Hanover, N.H., Sept. 1967.
13. Estrada, H., and Ward, S. R. "Forces Exerted by Ice on Ships," *Journal of Ship Research*, SNAME, Dec. 1968.
14. Chellis, R. D. *Pile Foundations*, McGraw-Hill, New York, 1961.
15. Peyton, H. R. "Ice and Marine Structures," 3 parts, *Ocean Industry*, Houston, March, Sept., and Dec. 1968.
16. Crighton, L. J. Icebreakers – Their Design and Construction, *Lloyd's Register of Shipping*, No. 45, London, 1959.
17. Zubov, N. N. *Arctic Ice*, U. S. Navy Electronics Lab., (AD 426972), San Diego, 1963.
18. Freiberger, A., and Lacks, H. "Ice-phobic Coatings for Deicing Naval Vessels," *Proc. 5th Naval Sciences Symposium*, 1961.
19. U. S. Army. *Shore Protection Manual*, Vol. 2, USACE, CERC, 1977.
20. Gerritsen, F. "The Effects of Ice on Hydraulic Structures," *Topics in Ocean Engineering*, Vol. 3, Gulf Publishing Co., Houston, 1976.
21. Brown, C. B. "AIDJEX Results on Arctic Ice Mechanics," *Proc. ASCE*, ST-2, Vol. 104, Feb. 1978.
22. Bercha, F. G., and Stenning, D. G. "Arctic Offshore Deepwater Ice-Structure Interactions," *Proc. OTC*, No. 3632, Houston, May 1979.

23.   American Association of State Highway and Transportation Officials, *Standard Specifications for Highway Bridges*, AASHTO, Washington D.C., 1977 and Interim 1978.

24.   Wortley, C. A. "Ice Engineering Guide for Design and Construction of Small Craft Harbors," Advisory Report #Wis-SG-78-417, University of Wisconsin Sea Grant, Madison, Wis., May 1978.

25.   Ibid. "Design of Harbor Structures for Boats," *Proc. Int. Conf.* on Port and Ocean Eng'g. under Arctic Conditions, Aug. 1979.

26.   "Ice Pressure Against Dams – A Symposium," *Trans. ASCE*, No. 2656, Vol. 119, 1954.

27.   Rose, E. "Thrust Exerted by Expanding Ice Sheet," *Trans. ASCE*, No. 2314, Vol. 112, 1947.

28.   Drouin, M., and Michel, B. *Pressure of Thermal Origin Exerted by Ice Sheets Upon Hydraulic Structures*, USACE, Cold Regions Research and Engineering Lab., Hanover, N.H., 1974.

29.   American Petroleum Institute. *Recommended Practice for Planning, Designing, and Constructing Fixed Offshore Platforms*, API, RP-2A, Dallas, March 1979.

30.   Nevel, D. E. "A Semi-Infinite Plate on an Elastic Foundation," Res. Report 136, U. S. Army Material Comm., Cold Regions Res. and Eng'g. Lab., Hanover, N. H., March 1965.

31.   Hetenyi, M. *Beams on Elastic Foundation*, University of Michigan Press, Ann Arbor, 1971.

32.   Michel, B. *Ice Pressure on Engineering Structures*, Cold Regions Science and Eng'g. Monograph III-B1b, USACE, CRREL, Hanover, N. H., June 1970.

33.   Voelker, R. P., and Levine, G. H. "Use of Ice Model Basins to Simulate and Predict Full Scale Ice-Structure Interactions," *Proc. OTC*, paper No. 1681, Houston, May 1972.

7
Deterioration of
Structures in
the Sea

The ocean environment presents a unique challenge to the structural engineer in terms of proper selection of materials for a given structure type and function over a range of local climatic and oceanographic conditions. Different materials are affected in various ways by the marine environment, the most notable effects including the corrosion of metals, spalling and degradation of concrete, attack of timber by marine organisms, and fouling and encrustation of virtually all materials. This chapter deals with the important ways in which the marine environment affects material durability and hence a structure's longevity and function. The discussion is oriented primarily about traditional civil engineering materials — steel, concrete and timber. Factors affecting all materials, such as marine growth, fouling, and general wear due to the incessant motion of the sea, are also discussed. Some relevant aspects of the physical and chemical environment are reviewed because an understanding of fundamental oceanographic processes is important in recognizing particular problems at specific sites. The spatial and temporal distribution of various deteriorating agents is emphasized.

7.1 INTRODUCTION

Sea structures are subject to various deteriorating agents throughout their lifetime, the degree of deterioration depending upon the climate, the properties of the seawater and its seasonal variations, tide range, and the type of material in regard to the climatic conditions and with respect to its relative immersion (i.e., splash zone, tidal zone, or continually immersed). Figure 7.1 illustrates the general

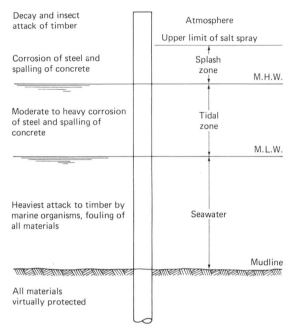

Fig. 7.1. Deterioration zones.

effects of relative immersion on various materials with regard to vertical zonation.

The marine atmosphere tends always to contain some small amount of salt, which increases the atmospheric rate of corrosion of marine structures over that of land structures. In timber structures above the splash zone, freshwater may collect and stagnate, initiating rot. The splash zone constitutes an area from the high water level to the upper levels attained by spray, and this zone is subjected to intermittent wetting and drying as waves run up or break on the structure. The tidal zone is the usual range between high and low water, which is periodically immersed. Below low tide to the seabed structures are continually immersed, and this is typically a zone of moderate to light attack of most materials except wood. Below the mudline most materials are relatively protected as the lack of oxygen prohibits oxidation and the existence of organisms.

Concrete, steel, and timber are the primary construction materials used in the sea as they are on land; so most of this discussion is devoted to them. Special materials of particular use in structural design such as stainless steel and other alloys and plastic and rubber compounds are mentioned in passing with respect to their particular structural application. Note that, in general, steel and concrete suffer the heaviest attack in the splash zone and the tidal zone, respectively, whereas timber is attacked below low water by marine organisms and is subject to decay above the splash zone. The important mechanisms by which timber, steel, and concrete structures are attacked will be discussed in the following sections of this chapter.

The relation of material durability to climate will become apparent, and disregarding other factors (e.g., cost, availability, ease of fabrication, etc.), it will be seen, for example, that steel structures are particularly vulnerable in the tropics, especially where there is frequent spray due to persistent wave action, and that concrete should be used with caution in freezing climates, especially with large tide ranges, owing to freeze–thaw action and subsequent spalling of the concrete.

Because the seas are in constant motion, chafe and wear are important considerations in designing for the marine environment. For example, chafe plates or rubbing strips should always be provided where mooring lines pass over any part of a structure, and structural elements subject to relative movement and mooring chains should be designed with extra metal thickness to account for wear. Fatigue of metal structures is a related problem that was discussed briefly in Section 3.7.

The discussion in Section 7.2 is devoted to some basic properties of seawater and general aspects of the marine environment, and to important oceanographic measurements that relate to the deterioration of structures in the sea.

7.2 ENVIRONMENTAL FACTORS

Physical and Chemical Factors

Among the most important basic concepts common to all aspects of oceanography is the relationship between temperature, salinity, and density of seawater. Salinity can be considered as the amount of dissolved solids in parts per thousand (‰) by weight in a water sample. Density typically increases with decreasing temperature and with increasing salinity. The interrelationships of temperature, density, and salinity are affected by pressure and, therefore, vary with depth. Oceanic temperatures typically vary from about 29°F at the poles and at great depths to perhaps 85°F for surface water near the equator, but most ocean water remains at 30° to 40°F below the surface, while surface values in the mid-latitudes vary from 30° to 60°F. The surface water down to about 10 to 20 feet is at a nearly uniform temperature. Then there is a very sharp decrease over the next 500 to 1000 feet, called the thermocline layer (a layer of mixing beyond which the temperature is fairly uniform or decreases slightly with depth). This pattern is generally typical of the vertical distribution of temperature with depth. Locally, however, there may be many anomalies and stratifications, especially where currents of separate water masses meet. The density of seawater is not very much different from that of freshwater, its mean specific gravity (S.G.) being about 1.026. Density is often given in terms of the density function (σ_t), which is equal to the specific gravity minus one, times one thousand or: $\sigma_t = (S.G. - 1) \times 1000$. Figure 7.2 shows the general relationship between temperature, salinity, and density. The mean salinity of the oceans is about 35‰ and typically varies from 31‰ to 38‰; however, for nearshore coastal waters it

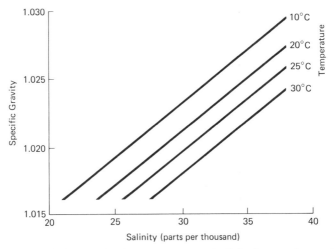

Fig. 7.2. Relationship between density, temperature, and salinity for normal sea surface water.

may be considerably less because of freshwater run-off, especially in the vicinity of large river mouths.

Water is considered an incompressible fluid for most engineering purposes. Were this exactly true, the oceans would be some 90 feet higher than the present sea level. The bulk modulus of elasticity of seawater at the surface is about 320,000 pounds per square inch (psi) at 37°F, and it does not vary significantly over the range of temperature encountered in the oceans. The error due to neglecting compressibility in pressure calculations amounts to about 2% at 30,000 feet of depth, so that for all practical purposes of structural design it can be safely neglected at depths up to several thousand feet. Some other mechanical properties of seawater are shown in Appendix 2.

The heat capacity of water is among the highest of all liquids and solids. This property explains the lack of extreme temperatures in the oceans and is most important to life in the sea in maintaining uniform body temperatures. The high heat capacity of water also permits large transfers of heat energy by water movement, thus drastically altering climates around the world. The latent heat of evaporation of water (the amount of heat required to evaporate a given amount of water) is the highest of all substances, a most important factor in the transfer of heat and water between the ocean and the atmosphere. Although the thermal conductivity of water is the highest of all liquids, it is still relatively low compared to many other substances. The thermal conductivity of seawater is slightly less than that of freshwater, and it increases with increasing temperature and pressure. Since the oceans are in constant motion and because turbulence greatly affects the conduction of heat, the coefficient of thermal conductivity used in most calculations of conductive heat transfer is the eddy coefficient. The eddy coefficient is many times larger than the coefficient of thermal conductivity for the still liquid and varies greatly with the degree of turbulence.

Light in the sea is attenuated by angular spreading, scattering by water molecules and suspended matter including organisms, and absorption by the water and suspended material. The extinction coefficient (β) is a measure of the decrease of light energy with depth. The longer wavelengths at the red end of the electromagnetic spectrum are absorbed near the surface. Because the shorter wavelengths are more effectively scattered and penetrate more deeply, the ocean generally appears blue. Coastal water is generally more turbid; that is, it contains more suspended material than oceanic water, so that light does not penetrate as deeply. Coastal water also often has a greenish-yellow tint, which is caused by minute plants and organisms suspended in the surface layers of the sea. Water color is recorded qualitatively for practical purposes on the Forel scale, which is accomplished by comparing the in situ water with standardized vials of various shades. Underwater vision is usually not possible below 100 feet, although greater depths are occasionally reported, and most of the time in the coastal environment vision is limited to 10 feet or less because of turbidity.

Seawater can be thought of as a dilute chemical soup containing nearly all the elements in solution in some amount. Elements in solution are found as ions, many of which combine and precipitate out as salts upon evaporation of the water. Elements can be grouped for convenience into three categories denoting their relative importance: the major elements, which are the most plentiful; the nutrient elements, which are essential to life processes; and the minor or trace elements, found in quantities typically less than one part per million (ppm). Concentrations are typically given in parts per million by weight, which are equivalent to milligrams per liter (mg/L). Table 7.1 lists some of the more important elements normally found in seawater at 35 ‰ salinity. There are 11 major elements, of which chlorine in the form of chloride ions (Cl^-) and sodium in the form of sodium ions (Na^+) are by far the most abundant. The remaining nine, in ionic form, are, in order of decreasing quantity: sulfur as sulfate (SO_4^{2-}), magnesium (Mg^{2+}), calcium (Ca^{2+}), potassium (K^+), carbon as bicarbonate (HCO_3^-), bromide (Br^-), strontium (Sr^+), boron (H_3BO_3), and fluoride (F^-). The ratios of these major ions do not vary significantly over the range of salinities normally observed in the oceans. Known as the law of constancy of major ion ratios, this relationship is most important to the determination of salinity and, hence, is central to many oceanographic calculations.

Salinity has already been defined as the total amount of dissolved solids in water. Given that the law of constancy of major ions holds, then if we could measure the concentration of any particular ion, we could obtain the total salinity by multiplying it by a constant.

Table 7.1. The composition of sea water at 35 ‰ salinity
(in order of ascending atomic weight).

Element	Micrograms per liter	Element	Micrograms per liter
Hydrogen	1.10×10^8	Molybdenum	10
Helium	0.0072	Ruthenium	0.0007
Lithium	170	Rhodium	
Beryllium	0.0006	Palladium	
Boron	4,450	Silver	0.28
Carbon (inorganic)	28,000	Cadmium	0.11
(dissoloved organic)	500	Indium	
Nitrogen (dissolved N_2)	15,500	Tin	0.81
(as NO_3^-, NO_2^-, NH_4^+)	670	Antimony	0.33
Oxygen (dissolved O_2)	6,000	Tellurium	
(as H_2O)	8.83×10^8	Iodine	64.
Fluorine	1300	Xenon	0.047
Neon	0.120	Cesium	0.30
Sodium	1.08×10^7	Barium	21
Magnesium	1.29×10^6	Lanthanum	0.0029
Aluminum	1	Cerium	0.0012
Silicon	2900	Praesodymium	0.00064
Phosphorus	88	Neodymium	0.0028
Sulfur	9.04×10^5	Samarium	0.00045
Chlorine	1.94×10^7	Europium	0.0013
Argon	450	Gadolinium	0.00070
Potassium	3.92×10^5	Terbium	0.00014
Calcium	4.11×10^5	Dysprosium	0.00091
Scandium	<0.004	Holmium	0.00022
Titanium	1	Erbium	0.00087
Vanadium	1.9	Thulium	0.00017
Chromium	0.2	Ytterbium	0.00082
Manganese	0.4	Lutetium	0.00015
Iron	3.4	Hafnium	<0.008
Cobalt	0.39	Tantalum	<0.0025
Nickel	6.6	Tungsten	<0.001
Copper	0.9	Rhenium	0.0084
Zinc	5	Osmium	
Gallium	0.03	Iridium	
Germanium	0.06	Platinum	
Arsenic	2.6	Gold	0.011
Selenium	0.090	Mercury	0.15
Bromine	6.73×10^4	Thallium	
Krypton	0.21	Lead	0.03
Rubidium	120	Bismuth	0.02
Strontium	8,100	Radium	1×10^{-7}
Yttrium	0.013	Thorium	0.0004
Zirconium	0.026	Protactinium	2×10^{-10}
Niobium	0.015	Uranium	3.3

This is what in fact is typically done by chemical oceanographers. Since the chloride ion concentration (chlorinity) is relatively easy to measure, salinity (S) is usually figured from chlorinity by the following formula:

$$S (‰) = .03 + 1.805 \text{ Chlorinity } (‰)$$

Ocean scientists employ a more rigorous definition of salinity for use in more exacting work.

The primary nutrient elements include carbon, oxygen, nitrogen, and phosphorous, which are essential to life processes. Silicon as silicate and calcium are utilized by many marine plants and animals in the construction of their cell walls or protective shells. Although silicon and calcium along with most of the trace elements are utilized in varying small amounts by organisms, they are not considered nutrients.

Carbon is the basic building block of life; it is fixed from carbon dioxide by plants in accordance with the following simplified chemical equation:

$$6CO_2 + 6H_2O \quad \xrightarrow[\substack{\text{plant} \\ \text{pigments}}]{\text{light}} \quad 6O_2 + C_6H_{12}O_6$$

This process, which is known as photosynthesis, requires as a catalyst light and the presence of plant pigments such as chlorophyll. Photosynthesis yields free oxygen and a sugar, such as glucose, shown in the above reaction. Sugars (carbohydrates) other than glucose may also be produced via photosynthesis. Oxygen is essential to respiration and is almost invariably found in one form or another in most organic compounds. Oxygen in the form of dissolved gas is most important in the corrosion and fouling history of a structure.

Nitrogen and phosphorus are continually recycled in the sea, and measurements of their concentrations and distribution are indications of biological activity. Since they are utilized primarily in the surface layers of the sea, their replenishment depends upon convective mixing and eddy diffusion.

Although the trace elements are generally found in minute concentrations in the sea, they are concentrated in the bodies of marine organisms. Copper, zinc, cobalt, iron, and molybdenum are especially important to many organisms. Unlike the major ions, concentrations of the trace elements may vary significantly locally because of river run-off, industrial pollution, and geological and biological processes.

Chemical reactions in the sea are governed by such chemical factors as acidity and alkalinity as given by the pH, the oxidation-reduction potential, and solubility.

The pH is the negative logarithm (base 10) of the hydrogen ion concentration:

$$pH = -\log_{10} (H^+)$$

Water molecules dissociate to form hydrogen ions (H^+) and hydroxyl ions (OH^-). The H^+ concentration of pure water is 10^{-7}; so its pH = 7. The pH scale ranges from 0 to 14, a pH less than 7 being considered acidic and greater than 7 being basic. Seawater is slightly alkaline, the pH of ocean water varying typically from 8.0 to 8.4.

The regulation of pH is important to life in the sea because most organisms cannot tolerate great changes in pH. Fortunately seawater is a buffered solution, which resists changes in pH. This buffering action works primarily through the mechanism of the carbon dioxide (CO_2)-carbonate (CO_3) system. When CO_2 is dissolved in seawater, primarily via its partial pressure in the atmosphere, it forms carbonic acid (H_2CO_3), thus lowering the pH. The H_2CO_3 in turn reacts with calcium carbonate ($CaCO_3$), raising the pH by the formation of bicarbonate ions (HCO^{3-}). This process, much simplified, is summarized as:

$$H_2CO_3 + CaCO_3 \longrightarrow Ca^{2+} + 2HCO_3^-$$

The actual maintenance of pH in the sea is much more complicated because it involves many factors other than the CO_2–CO_3 system. Local measurements of pH are important to ocean engineers in assessing the degree of biological activity and potential corrosion effects on structures.

Oxidation–reduction refers to the loss and gain of electrons, respectively, from the outer orbital shells of the reacting atoms.

The classic example, of great interest to designers of steel structures, is the formation of rust (geothite or hematite) when iron or steel is exposed to an oxidizing environment. Such a chemical reaction can be considered the sum of two half reactions: the loss of electrons (oxidation) of the iron and the subsequent gain of electrons (reduction) by the oxygen. The ions thus formed then combine to form the mineral geothite or hematite. Because there is a transfer of electrons we can consider that a voltage potential difference exists for the half cell reaction, and further that the reaction continues until an equilibrium condition is reached. At equilibrium the oxidation potential would equal the reduction potential. In seawater, the oxidation–reduction potential (E_h) is governed by:

$$2H_2O = O_2 + 4H^+ + 4 \text{ electrons}$$

All reactions in the open sea are controlled by this half cell reaction. Because the partial pressure of O_2 in the atmosphere and the hydrogen ion concentration are both fixed for the open sea, the E_h is also fixed. At a pH of about 8 and an O_2 partial pressure of about 0.2 atm, the E_h of seawater is about 0.75 volt.

Gases are absorbed in the ocean in direct proportion to their partial pressures in the atmosphere (Henry's law) and in accordance with an absorption coefficient for the given gas. Hence, all the gases found in the atmosphere are also found in the sea. The gases of primary interest are O_2 and CO_2 because of their importance in life processes, and perhaps nitrogen (N_2) because of its great abundance. Gas solubility increases with decreasing temperature, decreasing salinity, and increasing pressure.

Ocean water is typically saturated with gaseous nitrogen at all depths. Its concentration typically varies between 8 and 18 ml/liter of seawater.

The CO_2 concentration in the sea is the highest of all gases, the range being about 35 to 60 ml/liter. The CO_2 concentration found in the sea is not in accordance with Henry's law, as it enters into chemical reactions with both the water and some of the dissolved ions. Carbon dioxide is utilized by plants during photosynthesis and given off by animals as an end product of respiration.

Dissolved O_2 is of primary interest to ocean engineers as it affects the corrosion history of a structure and is an indication of the level

Table 7.2. Saturation values of oxygen in seawater (ml/L)*
from normal dry atmosphere (Fox, 1907).

Chlorinity (‰) Salinity (‰) Temperature (°C)	15 27.11	16 28.91	17 30.72	18 32.52	19 34.33	20 36.11
–2.	9.01	8.89	8.76	8.64	8.52	8.39
0.	8.55	8.43	8.32	8.20	8.08	7.97
5.	7.56	7.46	7.36	7.26	7.16	7.07
10.	6.77	6.69	6.60	6.52	6.44	6.35
15.	6.14	6.07	6.00	5.93	5.86	5.79
20.	5.63	5.56	5.50	5.44	5.38	5.31
25.	5.17	5.12	5.06	5.00	4.95	4.86
30.	4.74	4.68	4.63	4.58	4.52	4.46

*mg-atoms of oxygen per liter = $0.08931 \times$ ml/L.

of biological activity. Normal concentrations found in the sea range from 5 to 10 ml/liter where not affected by biological processes. Saturation values are given in Table 7.2 for various temperatures and salinities. Note that cold water can hold more dissolved O_2 than warmer water, which may be an important factor in the corrosion history of structures in cold water locations. Phytoplankton blooms may increase the concentration of O_2 locally to a point of supersaturation.

Biological Factors

The designer of structures must naturally be concerned with life processes in the sea as they relate to the problems of fouling and boring organisms and physical/chemical changes in the water due to biological activity, which may drastically affect corrosion rates. Marine animals are typically "curious," and anything new in the underwater neighborhood will soon become a source of investigation by the local inhabitants. The author has recently become aware of harbor seals on the eastern New England coast nibbling at the styrofoam flotation logs of marina floats. Fish have been known to nibble at deep sea fiber mooring lines, most likely to get at the algae and small organisms attached to them.

Structures themselves may have a beneficial effect on the local ecology, as evidenced by the increased catches of fish near offshore oil rigs in the Gulf of Mexico. The structures provide a base for the attachment of algae, which support small grazing organisms, which in turn attract fish and larger organisms.

In the study of ocean life it is convenient to classify the marine environment into various zones or biomes. The two major zones are essentially the water itself (pelagic) and the bottom (benthic). We can further divide the pelagic life forms into the oceanic and neritic provinces. Most structures of concern herein are built over the continental shelves in the neritic province, or near shore in the intertidal zone. Bottom life in the tidal zone is referred to as littoral; in the neritic as sublittoral. The sublittoral zone exhibits creeping and burrowing forms, some bottom-dwelling (demersal) fish, and sessile (permanently attached) organisms. The littoral zone is relegated primarily to burrowing forms and some sessile forms. Pelagic life is either free-swimming (the nekton) or floating and drifting with the prevailing water movement (the plankton).

The plankton can be further divided into zoöplankton (animal forms) and phytoplankton (plants). The phytoplankton are known as the primary producers (autotrophs), as they are able to synthesize organic matter from inorganic material in the presence of light and vegetable pigments such as chlorophyll. This process of photosynthesis, described above, utilizes CO_2 and water to produce O_2 and sugar (carbohydrates). Some bacterial forms are capable of synthesizing organic matter without light (chemosynthesis), and are of major importance in returning nutrients to the food chain. Nearly all primary production occurs in the upper 50 meters or photic zone, which marks the maximum depth to which light can be effectively utilized by plants. The process of photosynthesis is limited only by the supply of nutrients in the photic zone, and under ideal conditions as much as a 300% increase in mass of the producing plants can occur in a day. Most of the phytoplankton are algae of various types, primarily microscopic yellow-green algae, known as diatoms, and some mobile forms known as dinoflagellates. The autotrophs are preyed upon by heterotrophic forms. Small planktonic grazing animals such as copepods, which account for the majority of zoöplankton, are known as primary heterotrophs, and they themselves form the base of the food pyramid for successively larger organisms. Because of inefficiency of metabolic processes, it takes about 10 pounds of the consumed organism to produce 1 pound of consumer. Therefore, typically the larger the species, the fewer the number of individuals.

The total mass of living tissue (called the biomass) per unit volume or area of seawater is known as the standing crop. Estimates of the standing crop are central to all ecological studies and are generally specified in terms of wet weight, dry weight, or weight of organic carbon. The productivity is the change in the standing crop per unit time, given, for example, in hours, days, weeks, or even years. Primary productivity refers to the autotrophic fixation of CO_2 by photosynthesis, secondary productivity to the production of herbivorous organisms, and tertiary productivity to that of carnivorous organisms feeding upon the herbivores.

The role of bacteria both in the plankton and the benthos is of paramount importance in the recycling of elements and organic material back into the food chain. Various nitrifying bacteria convert ammonia (NH_3) from excreta into nitrites (NO_2) and then to nitrates (NO_3), which are nutrients of major importance to plant life. Various sulfur bacteria act similarly upon hydrogen sulfide (H_2S), yielding sulfur in a form usable by plants and animals. The production of H_2S due to bacterial decomposition is of primary interest to the ocean engineer. In stagnant seas or basins (of which the Black Sea is the outstanding example) there is no mixing or recirculation of water from the surface to the bottom. Under these conditions, the dissolved O_2 is soon used up, anaerobic conditions prevail, and H_2S is produced by sulfate-reducing bacteria, a reaction that may result in greatly increased corrosion rates. The pH may drop to 7.0 or lower under these conditions.

Seawater generally contains some suspended organic matter, measurable on the average as 1.2 to 2.0 mg of carbon per liter, and composed of excreta and decomposing organisms.

In areas where there is an upward flow of water, known as upwelling, nutrients are carried from the bottom to the surface layers where they can be utilized by plants and organisms; these areas are typically high in productivity. Such conditions typically exist where strong currents meet upward-sloping bottom contours, such as the Grand Banks of Newfoundland, or where prevailing offshore winds cause underwater currents to return shoreward, as along the West Coast of the United States. The nutrients of primary importance are: nitrates, for the production of protein; phosphates, essential to energy transfer; and bicarbonates, as a source of carbon.

Life forms are unevenly distributed throughout the seas in a seemingly random manner, with a tendency to congregate in certain areas. The distribution of organisms is affected by such physical conditions as temperature, salinity, light, supply of nutrients, and transport by wind and currents. The distribution is further modified by reproduction rates and the interactions of species. Organisms exhibit various thresholds of tolerance to fluctuations in temperature, salinity, light, and so on. Those that tolerate larger ranges of temperature and salinity are called eurythermal and euryhaline, respectively, whereas those that are relatively intolerant are stenothermal and stenohaline. The majority of marine fish cannot live in water with a salinity of much less than 20‰, because of the osmotic relationships of their body fluids to the surrounding medium. Most fishes' body fluids are nearly isotonic or slightly hypotonic to the seawater; that is, they exert nearly the same osmotic pressure, so that there is a slight outward flow through the animal's semipermeable membranes. In times past, even large seagoing vessels would travel up rivers to anchor in freshwater for the purpose of killing boring and fouling organisms.

Organisms are not only distributed areally and spatially, but also temporally. Many small planktonic creatures, notably the copepods, undergo daily vertical migrations, surfacing at night and sinking during the day. When the population is very dense, they account for the so-called deep scattering layer (DSL), a false bottom or bounce-back shown on echo sounders. Figure 7.3 shows the relative production of phytoplankton and zooplankton in a year for a typical North

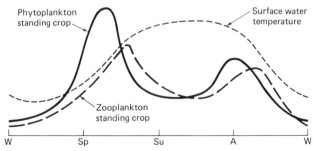

Fig. 7.3. Seasonal cycle of plankton growth (for temperate North Atlantic site).

Atlantic site. The diagram indicates blooms of phytoplankton in spring and autumn followed closely by an increase in the grazing zooplankton populations. Such temporal variations, caused by changes in water temperature, available light and nutrients, and so forth, are of interest in the study of the sequence and rate of fouling. Arctic waters typically show one bloom per year, in late winter or early spring, whereas tropical waters show many peaks, although overall production is lower. Overall production is highest in temperate and sub-Arctic waters.

Oceanographic Measurements

Oceanographers are concerned with gross measurements of the ocean's properties and characteristics. Such data are necessarily recorded to as high precision as possible. The ocean engineer, however, is typically concerned with local conditions, their extremes, and seasonal variations. This information generally need not be known to as high a degree of accuracy. The oceanographic data of prime interest to the structural designer with regard to the structure's durability include information from the following areas: physical/chemical properties of the water; sounding information, including water depth and the composition and character of the bottom; tide and current variations; sea and swell data; meteorological data; and ice conditions. Sea and swell, tide, current, and ice data required for design have already been discussed. The following discussion is intended to review briefly just what additional kinds of information on the marine environment are useful to the designer. Descriptions of oceanographic instruments and methods of data collection are not included. Interested readers should refer to references 1 and 2.

The physical/chemical properties of seawater most commonly recorded are temperature, salinity, dissolved oxygen, and pH. Because most fixed structures of concern here are built in relatively shallow water where the variation with depth is generally not dramatic, we are especially concerned with seasonal variations of the above properties.

Indicators of water quality, such as turbidity, and quantitative measurements of various pollutants may also be of interest to the structural designer. Figure 7.4 shows an environmental data report-

ENVIRONMENTAL DATA RECORD FORMAT

Date: Year:

		Jan	Feb	March	April	May	June	July	Aug	Sept	Oct	Nov	Dec
Tidal Range	mean high												
	mean low												
Tidal Current feet/second — maximum flood	surface												
	bottom												
maximum ebb	surface												
	bottom												
Temperature °C	surface												
	bottom												
Specific Gravity	surface												
	bottom												
Salinity 0/00	surface												
	bottom												
Dissolved Oxygen	surface												
	bottom												
Industrial Pollution	chemical												
	conc. ppm												
Turbidity (depth secchi disc)													
Biological Attack* (use attack rating from chart)	Limnoria												
	Teredinidae												
	Pholadidae												

Ratings for evaluating borer attack on test boards

Limnoria number of tunnels per square inch	Teredinidae number of tunnels per panel	Pholadidae number of tunnels per panel	Attack Rating*
one	up to 5	up to 5	Trace
10	6 to 25 (10% filled)	6 to 25 (10% filled)	Slight
25	26 to 100 (25% filled)	26 to 50 (40% filled)	Moderate
50	101 to 250 (50% filled)	51 to 100 (70% filled)	Medium Heavy
75	over 250 (75% filled)	101-200	Heavy
100 or more	filled, riddled, or destroyed	over 200, filled, or riddled	Very heavy

Fig. 7.4. Environmental data record format. (From U.S. Navy[3])

ing form reproduced from reference 3. Such a form is used for reporting data obtained in the study of marine borer activity. Turbidity is usually given in terms of the extinction coefficient (β), and is a measure of suspended solids. In itself, turbidity is not an especially useful measurement to the structural designer except that it indicates whether other tests should be made to determine the nature of the suspended matter. Industrial wastes, especially acids, are important in considering potential corrosion effects, whereas high nutrient concentrations mean increased fouling.

Nautical chart coverage is generally very good for assessing overall depths and bottom conditions for large areas. However, in siting a particular structure more accurate localized soundings are invariably required. All sounding information must be referenced to some tidal datum such as mean low water (see Section 5.1) and should include information on the character of the bottom. Underwater borings are taken in much the same way as land borings although generally with much more difficulty.

Annual and extreme temperature and precipitation at a given site are most important, especially in cold climates where ice and snow accumulation may be great. Very cold temperatures bring the threat of brittle fracture of metals, and frequent freezing threatens freeze-thaw damage to concrete structures. Temperature and rainfall records are generally available from local sources for most world locations. For offshore sites, information on ice conditions is required, as outlined in Section 6.1. General information and climatological data for all U.S. coastal waters, including the Great Lakes, Alaska, Hawaii, and Puerto Rico, can be found in the *U.S. Coast Pilots* issued by NOAA.[4]

7.3 FOULING

The word fouling refers in general to the accumulation of various plant growths and animal organisms on immersed and partially immersed surfaces. Depending upon the type and severity of fouling at a site, its effects should be considered in design. Some of the effects that fouling has on marine structures are as follows: increased drag, inertial, and gravity loads due to increased surface roughness, projected area, and mass; possible increase in corrosion rates due to destruction of protective coatings and oxygen concentration cell effects at the point of attachment of certain organisms such as barnacles; the abrasion and possible severance of mooring lines and cables; general bio-deterioration due to the direct deteriorating effects of certain organisms on certain materials; difficulty in inspection and maintenance due to the presence of prolific and tenacious growths. In the case of fixed and permanent structures, there is as yet not much a designer can do about fouling except to recognize that because of its ubiquitous nature it *will* occur. Therefore, drag and inertial force coefficients, projected areas, and so on, should be selected accordingly.

In the tropics, especially, hard mussel fouling may be a foot or more thick, and the apparent mass and projected area of a pile may be as much as doubled. It is likely, however, that during severe wave action at least the outer layers of fouling will be washed away. Based upon experimental studies,[5] it may be inferred that drag force coefficients may be increased on the order of 50% to 70% for light to medium fouling on a smooth 3-foot-diameter cylinder. Heaf[6] presented data on the effects of fouling on North Sea fixed platforms and concluded that marine growth can lead to significant increases in both overall and local loading on a given platform. A 2-inch thickness of growth led to an overall load increase of 5.5% on the typical platform studied. Heaf further notes that the consequences of fouling are more severe considering the effect on fatigue life compared to the design wave loading. In addition to considering increased loads due to fouling, the designer should also consider ease of cleaning and maintenance.

Fouling may be variously comprised of organisms from all the major groups, which include 2000 or more different species. The type and degree of fouling usually follows a given temporal sequence, which may vary with the season, water depth, changes in seawater properties, species abundance, and so on. Fouling is sometimes referred to as hard or soft fouling, for example, depending upon whether barnacles and hard-shelled organisms or algae and soft-bodied forms are predominant. Hard fouling may have a specific gravity of around 1.3 to 1.4, whereas that of soft fouling is around 1.0. Some of the more common and important forms of fouling are indicated in Fig. 7.5, which shows a typical vertical zonation of

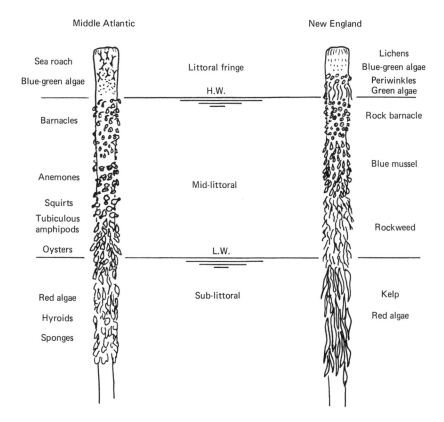

Fig. 7.5. Vertical Zonation of fouling communities. (Adapted from *A Field Guide to the Atlantic Seashore* by Kenneth L. Gosner. Houghton Mifflin Co. © 1978 by Kenneth L. Gosner.)

fouling organisms for two geographical locations on the Atlantic seacoast, as adapted from reference 7.

It is generally observed that bacteria and algal slimes are the first life forms to alight on a newly immersed surface, and it is believed that they form the base, or substrate, for the attachment of successive organisms. The usual pattern of bio-succession (i.e., the period of dominance of a given organism on a newly immersed surface) can be generalized as follows:[8] bacteria the first 1 to 3 days, algal slimes after 3 to 7 days, protozoans after 1 to 3 weeks, barnacles after 3 to 10 weeks, tunicates after 10 to 16 weeks, grasses after 3 to 6 months, and mussels after 6 to 12 months. Thereafter mussels may predominate, or some earlier organism may be dominant to the exclusion of mussels. The dominant organism ultimately depends upon geographic, climatic, physical/chemical, and biotic factors, etc. Because the reproductive rates of many organisms are regulated in large part by water temperatures, the degree of fouling may vary with the seasons in temperate climates (refer to Fig. 7.3), whereas it remains relatively constant in tropical climates. In general, the larger the organism, the longer the time fouling takes to become established.

The importance of bacteria in promoting fouling and in general bio-deterioration cannot be overemphasized. ZoBell[9] gave a detailed treatment of the subject of marine microorganisms, noting that bacteria are distributed throughout the water column but with their greatest concentration by far occurring at the seabed, where millions of cells per gram of wet mud may exist. Aerobic bacteria may occupy the upper inch or two of the bottom sediment, but the anaerobic bacteria are dominant below this level and are particularly important to the corrosion history of a structure, as they may exist under blisters of protective coatings and under the bases of sessile animals attached to metal structures. Sulfate-reducing bacteria may cause sulfide stress cracking of steels by producing H_2S under anaerobic conditions. Corrosion problems are discussed further in Section 7.5. In general, bacteria afford a foothold and food source for animals, and may aid the attachment of sessile organisms such as barnacles by promoting the deposition of calcareous cements. Fiber ropes are attacked by celluose-decomposing bacteria. Rubber products may be decomposed by certain bacteria, although laboratory tests[10] have shown polyethylene and neoprene to be relatively inert. Concrete sewer pipe and buried iron pipe have been severely damaged[10] by sulfur bacteria (*Thiobacillus*). The bacterium *Actinomyces* decomposes petroleum products and may be responsible for breaking down the oil-based treatment of timber piles, thus exposing them to attack by boring organisms.

Many important foulers, such as barnacles, are filter feeders, so that they require some degree of water motion to bring fresh nutrients and oxygen, and to remove wastes. Such organisms favor areas of significant tidal flow. Many organisms, again the barnacle is the prime example, exhibit definitive life stages, such as a free-living larval stage called a nauplius in the case of the barnacle, a settlement and initial attachment (cyprid stage) phase at which time the organism is quite small or even microscopic, and finally a growth and maturity stage. The mature barnacle consists of a glutinous integument with a calcareous shell made up of plates called scutes, inside of which are brushlike retractable feelers that filter food from the water. Extremely tenacious, barnacles can attach themselves in flowing water of up to perhaps 4 knots and remain attached at much higher water velocities, making themselves particularly troublesome in seawater intake structures. Barnacles can serve as a base of attachment for other organisms, and their shells remain permanently attached long after the organisms' death. The gluelike substance with which the barnacle attaches itself may eat through protective coatings. Oxygen concentration cells (as discussed in Section 7.5) may be set up about the base of a barnacle, resulting in heavy pitting corrosion.

Fouling prevention is a difficult problem. Although some metals exhibit a natural resistance to fouling,[11] they are exotic materials not feasible for general structural applications. Anti-fouling paints and inhibitors eventually must be reapplied, and thus are not of use on fixed and permanent structures. Electrical currents and resistive claddings have been used with limited success. Large offshore oil production platforms are cleaned by divers only at great expense. Benson et al.[12] have reported on various methods of fouling prevention. One novel approach to the problem was the introduction of grazing starfish on the mussel-covered legs of an oil production platform off the southern California coast.[13]

It has been reported[10] that the purple sea urchin (*Strongylocentrotus purpuratus*) was responsible for the demise of steel H-piles in 30 feet of water. The abrasive action of the urchins moving about over the piles' surface kept corrosion products cleared, thus continually exposing clean metal for further deterioration.

The extent of fouling may also be used as an indicator of the extent of marine borer activity.[14] Where only a few individual specimens of one or two species of fouling organisms are present, it is almost certain that marine borers will not be found.

General information on types and extent of fouling for many coastal areas of the world can be found in reference 15.

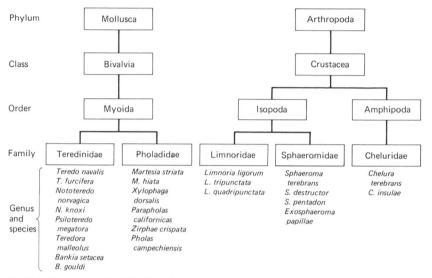

Fig. 7.6. Taxonomic classification of some important marine borers found in North American coastal waters.

7.4 MARINE BORERS AND DECAY

Certain marine organisms, primarily from the taxonomic groupings of molluscs and crustaceans, will bore into and destroy timber, soft concrete, and other materials immersed in seawater and in the tidal zone. Figure 7.6 shows the taxonomic classification of some of the more common and destructive marine borers active in U.S. coastal waters and found also in other parts of the world. Two families, the Teredinidae and the Pholadidae, contain most of the damaging species from the class Bivalvia (pelecypods), phylum Mollusca; and three families, the Limnoridae, Sphaeromidae, and Cheluridae of the class Crustacea, phylum Arthropoda, complete the list of the most important boring organisms. Certain insects, also from the phylum Arthropoda, destroy timber above the high water level and will be discussed at the end of this section. Figure 7.7 illustrates marine borer damage to a section of marine railway trackage constructed of treated and sheathed timber exposed for only a few years in Florida coastal waters.

The teredos, commonly known as shipworms, are perhaps the most treacherous of marine borers because the extent of their

Fig. 7.7. Marine borer damage to treated timber track of marine railway removed after only a few years immersion in Florida's Atlantic coastal waters. *(Photo courtesy of Crandall Dry Dock Engineers, Inc.)*

damage remains largely unknown until the structure becomes unserviceable or totally collapses. Figure 7.8 shows a cross section of a timber pile exposed only 16 months in Costa Rican coastal waters. A diagrammatic representation of one of the most common species, *Teredo navalis*, is shown in Fig. 7.9. The adult teredo somewhat resembles a worm with two clamlike shells at its head end and two siphons and a structure called a pallet at its tail end. Figure 7.10 shows a live specimen embedded in a piece of soft wood. The animal enters wood in a larval form and leaves not much more than a pinhole visible at the wood's surface. The animal matures inside the wood tunneling, usually but not necessarily downward, as it grows, using its rasplike shells to cut through the wood and ingesting the wood fiber (cellulose). The animal also circulates seawater through its body via incurrent and excurrent siphons, which are exposed at the surface. However, when the animal is disturbed by either physical trauma or changing environmental conditions, the siphons are withdrawn, and the end of the burrow is blocked by the pallet. The

Fig. 7.8. Teredo damage to pile taken from wharf at Estero Azul, Costa Rica after only 16 months in place. Section was taken at M.L.W.. *(Photo courtesy of Battelles W.F. Clapp Laboratories.)*

ability to isolate itself via this mechanism is very important to the survival and distribution of the species, as the teredo can live sealed inside its home for weeks, even when the wood is removed from the water. In this way the animal can survive temporary changes in water temperature and salinity that would otherwise be lethal, and it can travel across oceans borne by currents or wooden vessels, ready to infest new areas not previously known to have borers. Adult teredos are typically on the order of 6 to 12 inches long and perhaps up to one-half inch in diameter, while individuals up to 6 feet long and over 1 inch in diameter have been reported.[16]   Discussion of the complete anatomy and natural history of the Teredinidae can be

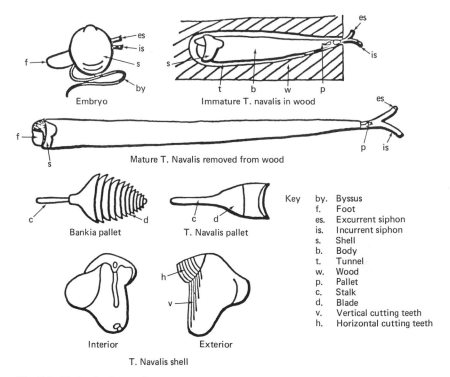

Fig. 7.9. Marine boring molluscs. (From *Marine Biology Operational Handbook,* Navdocks MO-311[3])

found in the work of Turner.[17]   Teredos are probably the most cosmopolitan of the borers, although both the teredos and *Limnoria* are distributed worldwide.   On the West Coast of North America, *Bankia setacea*, a close relative of the teredo, is particularly destructive; it is well distributed in the Pacific Northwest because of the many drifting logs moved about by the logging industry.

The pholads, known as rock borers, resemble ordinary clams somewhat, as their bodies are entirely inside their shells.   These animals may bore into rock and porous concrete as well as into timber.   They typically make shallow oval-shaped burrows at the timber's surface.   The genus *Martesia*, probably the most prominent boring pholad, has been known to penetrate the solid lead sheathing of underwater power cables.[10]   Other members of the family Pholadidae reportedly attacked the concrete jackets on timber piles in

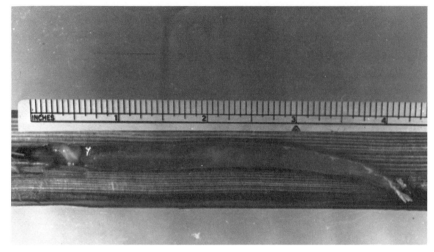

Fig. 7.10. Adult Teredo inside wood tunnel. *(Photo courtesy of Battelles W.F. Clapp Laboratories.)*

Los Angeles harbor.[10]   Seven to eight animals per square foot were found inhabitating 16 of 18 piles.   The concrete in this case was relatively weak, consisting of cement mortar with no coarse aggregate.

The mode of attack of the crustacean borers is somewhat different from that of the molluscs (refer to Fig. 7.11).   The limnorians, for example, excavate shallow burrows and tunnels at the timber sur-face, most commonly between the mudline and the low water level. The thin walls of their burrows are easily washed away, thus expos-ing new wood for attack.   The burrowing density of *Limnoria* may be up to 200 to 300 burrows per square inch, and they may reduce a timber pile diameter by 2 inches or more per year.[16]   *Limnoria* is commonly known as the woodgribble, and is typically 1/8″ to 3/16″ in length.   *Limnoria* burrows often also contain the *Chelura*, a related borer, which is somewhat less active than the limnorians. Members of the genus *Sphaeroma*, known as the pillbug, are similar to but larger than the *Limnoria*, with more rounded oval-shaped bodies, and are particularly active in European and Mediterranean waters.   An enlarged view of *Limnoria* in action is shown in Fig. 7.12.

All types of borers respond to differences in environmental conditions.   Borers are generally more active in the tropics and with-in the euphotic zone and within normal oceanic ranges of salinity.

Fig. 7.11. Limnoria Lignorum in wood showing burrows. *(Photo courtesy of Battelles W.F. Clapp Laboratories.)*

Ninety percent of the major borer types possess a larval stage that is 20% to 100% more susceptible to toxins and adverse conditions than the adult stage. Borers burrow for food or protection or both, and all borers depend upon the chemical dissolution of calcium carbonate ($CaCO_3$) and some mechanism of either rasping and wearing or chewing and ingestion.[18] The breeding period of most borers is largely regulated by water temperature; hence the general decrease in activity with increase in latitude. Almost all borers (except a few rare freshwater varieties) are inactive in salinities less than

Fig. 7.12. Close up view of Limnoria at work in burrows. *(Photo courtesy of Battelles W.F. Clapp Laboratories.)*

about 10 parts per thousand (‰), and generally activity is much restricted with salinities below 15‰. All borers are active in salinities above 20‰. *Teredo* can tolerate greater ranges and fluctuation of salinity, down to perhaps 8‰, that would be lethal to *Limnoria*, which is generally inactive below salinities of 12‰ to 16‰. Within a family group certain species may be dominant within specific environmental conditions; for example, in the family Limnoridae the optimum temperature ranges for three genera are: 5° to 10°C for *L. lignorum*, 12° to 16°C for *L. quadrapunctata*, and > 16° to 20° for *L. tripunctata*, although they may survive temperatures as low as 0°C. Borer activity is generally decreased by increased turbidity and siltation, by increased light, and by fast-moving currents. *Teredo* will not attack in currents in excess of about 1.4 knots and

Limnoria in currents over about 1.8 knots.[16]   The pH is especially critical and must fall in the range of 7.5 to 8.5 for survival.   The dissolved oxygen content should preferably be within normal limits for borers to be active.   The presence of hydrogen sulfide (a reducing environment) tends to lower the pH and hence curtail or halt borer activity.   At sites where few or no fouling organisms are present, it is unlikely that borers will be active.   No site should be considered to be free of borers, however, even if there is presently no apparent activity, as changes in environmental conditions and/or the introduction of borers by currents or vessels may infest a previously borer-free site.   Site-specific records of marine borer activity can be found for the U.S. East Coast and worldwide in the work of the William F. Clapp Laboratories,[19, 20] sponsored by the U.S. Navy, and for the U. S. West Coast in the bulletins of B. C. Research.[21]

The use of timber for the immersed portions of structures should be avoided in the tropics and at sites where borers are known to be voracious.   Wherever timber is used in seawater, it should at the least be treated.   The most common method is that of pressure treatment with creosote, usually to a retention of 16 to 24 pounds per cubic inch.   Not all woods will take pressure treatment.   Table 7.3, after Crandall,[22] shows the creosote retentions of some common structural woods.   Creosote treatment is not worth much in the tropics; here wood must be avoided or protected with sheathing and cladding, and/or some form of naturally resistant tropical hardwood should be used.   Greenheart (*Ocotea rodiaei*) is the most outstanding example of a naturally resistant wood; it apparently possesses certain chemical agents toxic to most borers.   Properties of greenheart and some other naturally resistant woods are given in Chellis,[16] which reference also contains a rigorous discussion of timber pile protection and properties.   Recently PVC and other plastic wrappings have been employed with apparent success, provided that the wrapping remained intact.

Above the high water level, moist wood is susceptible to decay through the action of bacteria, fungi (mycelium), and the grubs and larvae of certain beetles.   Wood-boring beetles are generally classed as round-headed or flat-headed borers, and most species are not exclusively marine.   The *Nacerda*, particularly *Narcerda melanura*, known as the wharf borer, have been especially damaging to wharves and bulkheads in the northeastern United States.   This organism has

Table 7.3. Creosote retention of common structural woods.

Woods with Good Retention	Uses
Yellow Pine – sap and summer wood only	Piling, Shims
Loblolly Pine	Sheathing, & Bracing
Jack Pine	Bracing, Sheathing & Piling
Red Oak (with careful treatment)	Fender Piles, Shims
Red Pine	
Woods with Light Retention	
Birch	Shims
Hemlock	Cribbing
Woods with No Retention	
Fir	Flooring, Beams & Piling
White Oak	Fenders Beams & Piling
Yellow Pine – heartwood	Beams & Piling
Spruce	Piling
Woods Resistant to Marine Borers Without Creosote	
Azobe	Beams
Manbarklak	Beams
Greenhart (temperate regions only)	Piling and Beams
Oak and Hemlock piling while bark stays on	
Angelique or Basra Locus	

After Crandall. [22]

a whitish grub about 1 inch long, which eats away timber, especially damp and already rotted pile tops and timbers. Rot, in general, is caused by fungi, which require air, water, food, and warmth in order to survive. Rot may manifest itself in a bleached, whitish, appearance of the wood's surface, or as a brown powder residue "dry rot" with the wood surface notably eaten away and punky. Isolated areas where freshwater can collect, like the checked top butt end of a timber pile, are likely places for rot to begin. Preventing freshwater from collecting, providing adequate ventilation, and using protective coatings will go a long way toward preventing rot.

Timber continually immersed and especially timber immersed in seawater not subjected to borer action will be preserved indefinitely, as will all timber below the mudline.

7.5 CORROSION

The subject of marine corrosion in general is well covered in several references, such as references 23–26, and a comprehensive discussion could fill many volumes. Therefore, only a cursory overview of corrosion problems peculiar to marine structures will be attempted here. The reader is referred to the literature for further enlightenment on this important subject.

Corrosion, in general, is due to an electrochemical process whereby atoms of a given metal lose electrons (oxidation), thus becoming positively charged. The free electrons then combine (reduction) with atoms of an adjacent area or surrounding substance. The migration of electrons (an electric current) from an anodic area to a cathodic area may be caused or accelerated by various conditions. Corrosion can be classified according to the conditions or type of electrochemical process, some of which processes are briefly described in the following paragraphs.

General and pitting corrosion is exhibited by the scaling or pitting of the surface of a metal from reaction with its surroundings or a difference in potential between different areas of the same metal. All metals form a thin oxide film when exposed to the atmosphere. This protective film is more or less easily broken down or washed away in various types of metals, resulting in further corrosion. Water, especially seawater or acidic water, serves as a transport medium (electrolyte) for electrons and thus accelerates the corrosion process. Alternate wetting and drying accelerate the corrosion process by washing away successive layers of oxide film.

Crevice corrosion is formed in isolated areas where less oxygen may be available for the repair of the oxide film than on an adjacent exposed surface; thus, a difference in potential is set up.

Selective corrosion, such as the dezincification of some brasses, is a process whereby one particular metal of an alloy is attacked, eventually destroying the metal's alloyed properties.

Stress-corrosion cracking is a process whereby normal corrosion is accelerated by tensile stresses within the material.

Corrosion fatigue occurs in cyclically loaded members or members subject to continual load reversals. The working of the material accelerates the corrosion process.

Galvanic corrosion is caused by a difference in electric potential of two dissimilar metals in direct contact or connected by a metallic path in a conducting medium such as seawater. Electric potentials of various metals in seawater are known (see Table 7.4); so galvanic

Table 7.4. Galvanic series of metals in quiet seawater with emphasis on structural steel.

		Metal	Voltage Potential (SCE)
(–) Anodic		Magnesium alloy	–1.6
			–1.4
			–1.2
		Zinc Aluminum alloy Galvanized iron	–1.0
			–0.8
		Mild steel (clean)	–0.6
	Lead Stainless steel (active)	Cast iron	
		Mild steel (corroded)	–0.4
	Brass	Mild steel (in concrete) Mill scale on steel	–0.2
	Copper Bronze	Stainless steel (passive)	0.0
			+0.2
		Graphite	
(+) Cathodic			+0.4

(Current flow arrow pointing downward on right side)

Note: Relative position of metals on voltage scale are approximate for average conditions but may vary considerably for given metal depending upon alloy content and other physio-chemical factors. Ranges of potential are shown for mild steel in order to emphasize the possibility of galvanic action on the same metal.

corrosion can generally be avoided by not using metals with large differences in electric potential in the same area. Note in Table 7.4 that, as a practical matter, corroded steel is higher on the galvanic scale (more "noble") than new steel. This fact is important when considering the replacement of corroded steel on an existing structure, as a patch of new steel welded to the existing will become anodic and thus exhibit accelerated corrosion. This effect can be offset somewhat by making the new steel thicker or more massive than the existing metal. In general, the greater the separation of metals in the galvanic series, the greater will be the rate of corrosion of the less noble metal.

Corrosion can be initiated by and accelerated by stray electric currents, various marine organisms such as barnacles, and exposure to wear such as the erosion-corrosion of steel sheeting by exposure to wave and sand action. Structural steel corrodes at an average rate of around .005 inch (.13 mm) per year in quiet seawater, but pit growth may be up to ten times as great.[27] By comparison, brasses and other copper alloys corrode at an average rate of .001 inch (.03 mm) per year under the same conditions. Moving water accelerates corrosion of all materials.

Corrosion of steel in seawater often follows a characteristic vertical distribution, which may, however, vary somewhat with geographical location. The rate of corrosion, for example, in a vertical pile is typically light to moderate in the upper part exposed only to the atmosphere, as only oxidation corrosion with atmospheric moisture takes place. In the so-called splash zone where the structure is intermittently wetted and dried and the protective film of corrosion is continually washed away, the rate of corrosion is the greatest, perhaps three to five times as great as for exposure only to the atmosphere. Just below the high tide mark corrosion often decreases to near a minimum, as this zone is somewhat protected by an oxygen concentration cell effect; however, in some structures other than vertical piles, corrosive attack may be heavy in this zone. The rate again increases in the continuously immersed portion of the pile, especially in the upper layers where there is plenty of dissolved oxygen or when there are strong currents to accelerate the galvanic effect. Below the mudline, corrosion becomes

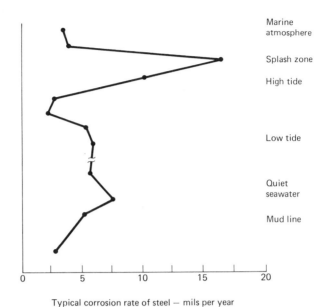

Marine
atmosphere

Splash zone

High tide

Low tide

Quiet
seawater

Mud line

0   5   10   15   20

Typical corrosion rate of steel — mils per year

Fig. 7.13. Corrosion of steel piling in seawater, generalized vertical profile. (From *Guidelines for Selection of Marine Materials,* International Nickel Company, Inc.[27])

minimal, as there is little or no oxygen present, and protective films remain intact. Figure 7.13, from reference 27, shows this characteristic pattern for a steel pile at Kure Beach, North Carolina.

Ayers and Stokes[28] presented vertical profiles of corrosion rates of steel sheet pile bulkheads based upon field studies conducted at eight widely scattered U. S. naval stations. Overall corrosion rates and the zones of maximum attack varied with climate, tide range, and other factors. For example, at Boston the highest rates occurred at the mudline and 1 foot above M.L.W. and were approximately .008″/year, or 8 mils per year (MPY). By contrast, at Norfolk, Virginia maximum rates of about 17 MPY occurred about 2 feet below M.L.W. At Key West, Florida the maximum rate observed was nearly 20 MPY about 2 feet above M.H.W. These sites, in order of introduction, represent a transition from colder to warmer climates and higher to lower tide ranges. Maximum rates for all three locations were on the order of twice the average rates. Rates for other stations are summarized in Table 7.5, reproduced from reference 29 after Ayers and Stokes. Environmental factors affecting marine corrosion rates in general are summarized in Fig. 7.14.

Table 7.5. Corrosion of Steel Piles.

Station	Maximum corrosion rate (mils per yr.)					Remarks
	Mud line	M.L.W.	Mid tide	M.H.W.		
Boston Naval Shipyard, Boston, Mass.	7.9	8.2 (+1 ft.)	......	......		Shallow water, 5 ft. max. depth, 10 ft. tide range.
Norfolk Naval Shipyard, Portsmouth, Va.	......	17 (−2 ft.)	......	......		Bitumastic coating originally applied; little in evidence near m.l.w. line.
Key West Naval Station, Key West, Fla.	......	10.3 (−2 ft.)	......	19 (+2 ft.)		Coated full length.
U. S. Naval Station, Coco Solo, Canal Zone.	4.0 (−1 ft.)	8.8 (−2 ft.)	......	17.3 (+2 ft.)		
Puget Sound Naval Shipyard, Bremerton, Wash.	......	8.5 (−2 ft.)	......	10 (splash zone)		No protective coating. Cathodic protection added below m.h.w. level.
U. S. Naval Air Station, Alemeda, Calif.	......	......	......	8.6 (2/3 of M.L.W. to M.H.W.)		Film of fuel oil coats piling.
U. S. Naval Station, San Diego, Calif...	......	14 (+1 ft.)	......	14 (+2 ft.)		Sewage treatment discharge; little change of water.
Pearl Harbor Naval Shipyard, Hawaii...	......	......	10	10 (+2 ft.), 11 (+6 ft.)		Bitumastic coating, no coating above half depth.

From U. S. Navy [29] after Ayers and Stokes. [28]

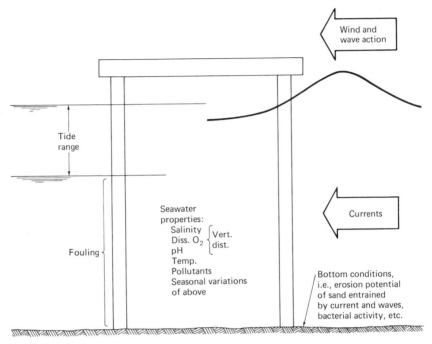

Fig. 7.14. Environmental factors affecting corrosion of marine structures.

As noted previously, biological activity, such as the action of sulfate-reducing bacteria in a reducing environment, can greatly accelerate the corrosion of metal structures. Certain organisms that attach themselves to or move over the surface of metal structures may cause pitting and erosion-corrosion. Anaerobic conditions at the mudline can cause greatly increased corrosion due to bacterial activity. In some instances, however, fouling may be beneficial, as in the case of a uniform coating of slime or weeds which may effectively insulate the steel surface from corrosion.

The frequency and the degree of wave action are important to the rate of splash zone corrosion, owing to the frequency and extent of alternate wetting and drying. Currents increase the rate of corrosion; in the case of mild steel the rate may be increased on the order of three to six times the quiescent rate for water velocities of 5 and 10 feet per second, respectively.[27]

Tide range affects the distribution of corrosion rates over the vertical extent of the structure, depending upon total water depth and exposure. In areas where the range is large, for example, the trend seems to be for the greatest attack to be near or just above mean low water, whereas in areas with small tide range the attack is greatest in the splash zone. The distribution of corrosion in general is due to zones of differential aeration that are set up.

Increases in temperature, salinity, and dissolved oxygen content all tend to increase corrosion rates. Consideration should be given to the seasonal variations in these factors, as corrosion rates may vary throughout the year.

Table 7.6, reproduced from reference 29, shows some typical corrosion rates for various types of steel sheet pile structures as affected by geographic location, tide range, earth cover, exposure, and protective coatings. This table is based upon studies conducted by the U. S. Army Corps of Engineers, Beach Erosion Board, as described by Rayner.[30] In general, harbor bulkheads exhibited much lower rates than beach bulkheads and groins and jetties situated in the same geographical location. The latter type structures are subjected to erosion-corrosion due to the abrasion effects of entrained sand and the greater exposure to turbulent wave action. This is also revealed in the fact that groins and jetties exhibited much higher rates of metal loss at mean tide and M.L.W. than at higher elevations observed for harbor and beach bulkheads. Northern locations had lower rates of loss than southern locations. Earth cover and painting materially reduced the rate of loss. Rayner gives the mean rate of loss of thickness of steel sheet piling, based upon 451 weighted averages, as about 8 MPY.

The engineer cannot assume that average corrosion rates for a given site will fall within some predetermined range. Although the rates and distribution of corrosion indicated in Tables 7.5 and 7.6 are representative, corrosion can be a highly localized phenomenon. Pitting can be many times the average rate, and splash zone rates as high as 55 MPY have been reported on offshore structures.[31] Typical corrosion rates of unprotected steel in offshore structures have been reported as: 0 to 10 MPY submerged, 25 to 40 MPY in the splash zone, and 3 to 5 MPY in the atmospheric zone.[16] High rates of corrosion are not necessarily relegated to warm climates, as strong currents and the high dissolved oxygen content of cold water can cause exceptionally high rates of corrosion.[32, 33] Corrosion can significantly affect the fatigue life of a structure.[34] There are nu-

Table 7.6. Average rates of deterioration of sheet pile structures.

Factors affecting deterioration		Types of structure		
		Harbor bulkheads (in. per yr.)	Beach bulkheads (in. per yr.)	Groins and jetties (in. per yr.)
Geographical location	Southern region	0.0062	0.017	0.018
	Northern region	0.0023	0.0075	0.011
Zone relative to tidal planes.	8 ft. above M.H.W.	0.0049	0.020	0.010
	5 ft. to 8 ft. above M.H.W.	0.0049	0.022	0.010
	2 ft. to 5 ft. above M.W.H.	0.0049	0.0081	0.010
	M.H.W.	0.0027	0.0074	0.0055
	Mean tide level	0.0024	0.001	0.024
	M.L.W.	0.0035	0.002	0.028
	M.L.W. to ground line	0.0035	0.0035	0.0035
	Below ground line.	0.002	0.002	0.002
Sand, earth or other cover	No cover on either surface of pile.	0.0075	0.027	0.019
	One surface never covered, other covered part time.	0.0076	0.020	0.014
	One surface only covered	0.0026	0.0094	0.020
	One surface always covered, other covered part time.	....	0.0065	0.0057
	Both surfaces covered part time.	....	....	0.017
	Both surfaces always covered	....	0.0017	0.0026
Exposure to salt spray	Heavy spray.	0.0083	0.016	0.016
	Moderate spray.	0.0041	....	
	Light spray.	0.0024	....	
Painting	None.	0.0045	0.018	0.020
	At least once.	0.0027	0.011	0.010

From U. S. Navy [29] after Rayner. [30]

merous articles and publications on the corrosion and corrosion protection of offshore structures and marine corrosion in general. The reader is referred in particular to the proceedings of the Offshore Technology Conference (OTC), held annually since 1969, and the periodicals and conference proceedings of the National Association of Corrosion Engineers (NACE).

Steel H-pile sections are in common use in waterfront construction because of their availability, ease of splicing and driving, and so on, but from a corrosion point of view they are clearly inferior to pipe piles of the same capacity. For example, most marine structures involve piles with relatively long unsupported lengths, so that considerations of elastic stability govern the allowable stresses. When metal is removed from the flanges, and often this is where corrosion rates are highest in H-piles,[16] the radius of gyration of the pile cross section may be drastically reduced as compared to the same amount of metal loss on an equivalent pipe pile. Pipe piles have the added advantage of being easier to maintain and inspect, and present a minimal surface area per weight of pile.

Structural steel is to date the metal used almost exclusively in the main structural elements for both waterfront and offshore structures; thus corrosion of other metals is not discussed herein. Because of fatigue, weldability, and general corrosion and econmic considerations, mild structural steel (carbon steel) and high strength, low alloy steels as used in building construction are probably the best all-around choice. Higher-strength steels are more susceptible to failure by corrosion fatigue, by brittle fracture in cold climates, and by hydrogen embrittlement. Some steels especially developed for marine use, alloyed with about .5% copper and .5% nickel, show better corrosion resistance in the splash zone, two to three times that of mild steel, and are generally recommended for sheet piling bulkheads.[35]

Studies conducted for the U.S. Navy[36] showed that the use of most alloys in structural steels had little effect on corrosion rates, but that in general wrought iron performed well with corrosion rates considerably less than that of the alloyed steel. Cast iron has an enviable record of corrosion resistance in seawater, although it is a brittle material.

Mill scale is highly resistant to corrosion and when intact will protect the base metal; but if the base metal is exposed by flaking off of the mill scale, greatly accelerated localized corrosion and pitting will result. Galvanizing of steel fastenings and hardware should be called for above low water. However, it is of little use below low water, as the zinc quickly goes into solution. Stainless steels are used for fastenings and are generally corrosion-resistant, but are susceptible to pitting. Stainless steels are particularly susceptible in low oxygen, or reducing, environments.

Wire rope is usually treacherous in salt water. It suffers from crevice corrosion between strands and may fail from axial or bending fatigue and abrasion. Stainless steel wire rope is subject to pitting at any zone of immersion, whereas galvanized mild (plow) steel rope is relatively serviceable in the splash zone where galvanizing remains intact. Chain, although cumbersome, is often a good substitute where flexible tension members are required. Chain has the additional advantage that it can be checked for its reserve strength by gaging the remaining metal thickness, whereas it is nearly impossible to determine the reserve strength of wire rope. Reference 37 summarizes the results of seawater immersion tests on corrosion rates, changes in mechanical properties, and so on, for various steels, alloys, anchor chain, and wire rope as reported by the U. S. Naval Civil Engineering Laboratory.

All permanent steel structures must have some form of corrosion protection regardless of location. Traditionally structures have been designed with extra metal thickness to account for the expected metal loss over the design life of the structure. This practice is not wholly satisfactory because local pit growth can be many times the expected average rate, and on large structures even 1/16-inch additional thickness of metal can amount to a considerable tonnage of steel. Similarly, the practice of using lower unit stresses in structural design is not wholly satisfactory from a purely corrosion point of view. There are myriad protective coatings on the market with all kinds of claims and all kinds of prices. Protective coatings and/or claddings are required in the splash zone and tidal zone. Below low water, however, cathodic protection systems should be considered. Chellis[16] thoroughly describes the various means of protecting pile structures in the marine environment, and Johnson[38] covers deterioration, maintenance, and repair of structures in general. Description of protective coating and cathodic protection systems is beyond the scope of this book; the reader is referred to the literature and to the brochures of the various manufacturers.

Corrosion can be minimized by functional design, proper selection of materials, protective coatings, claddings such as concrete encasements, rust inhibitors, and cathodic protection systems using impressed currents and sacrificial anodes. When dissimilar metals are used, a galvanic series table should be consulted to verify the potential difference between them. Functional design details would include such things as the avoidance of overlapping plates, discontinuous welds, and connection details that provide crevices or stress concentrations. The use of back-to-back angles or open-ended tube and pipe sections that cannot be maintained is tantamount to inviting future problems.

7.6 MARINE CONCRETE

Concrete when properly made and placed is an excellent material for ocean construction, whereas improperly made concrete will deteriorate rapidly. The chief enemies of concrete are alternate freezing and thawing in cold climates, abrasion, and chemical attack. The most important aspect of marine concrete is obtaining a dense and durable finished product. Permeability is probably the most important single property in this respect.

Chemical attack takes place primarily from the action of chlorides and sulfates in the seawater, a type of attack that proceeds most rapidly in warmer climates. The presence of $H_2 S$ in some locations may result in acid attack of the cement and, hence, accelerated deterioration. Corrosion of reinforcing steel occurs at a pH lower than 11, and since the mean pH of seawater is slightly greater than 8, it is important that the concrete cover maintains a relatively high pH for the steel.

Concrete that is continually submerged will usually perform quite well and gain strength with the passage of time. In the tidal zone, however, it is subject to abrasion and to freeze-thaw cycling in cold climates. Freeze–thaw is one of concrete's worst enemies,[39] and is a progressive deterioration that may repeat a cycle with every rise and fall of the tide.

Prestressing of concrete closes up cracks, offers relatively high strength and durability, and is generally desirable for applications in the marine environment.

The usual specification for structural concrete for ocean use, in all but massive structures, requires a maximum aggregate size of about 3/4 to 1 1/2 inch (18 mm), a high ratio of cement to sand and a low ratio of water to cement, and often the use of an air-entraining admixture (about 3% to 5% air by volume) to ensure durability. Air entrainment is essential in freezing climates.[39] It is also usual to call for a type II or III Portland cement, preferably type II for its sulfate-resistant properties. In addition, aggregates should be as nonreactive and abrasion-resistant as possible, and sharp corners or edges on structural elements should be avoided. Concrete actually poured in the water through a tremie pipe or lowered in individual sacks should also have a minimum amount of clay binder, such as bentonite, to keep the cement particles from leaching out of the mix. Concrete cures relatively rapidly under water because there is a constant supply of moisture to complete the hydration reaction. Proportioning of concrete mixes and structural design of concrete is beyond the scope of this book; the interested reader is referred to Gerwick,[40] who has presented a general description of the application of concrete in marine construction.

Adequate cover over the reinforcing steel must be provided because seawater will rapidly cause the reinforcement to corrode and, upon doing so, to expand, spalling the concrete. (See earlier comments on pH and steel corrosion.) The corrosion products of steel occupy roughly 12 to 13 times the volume of the uncorroded steel, and once corrosion of the steel begins it will exert a bursting effect on the concrete and progressive spalling will result.

Bazant[41] has presented a physical model for the corrosion of steel reinforcing, from which the corrosion rate and time to cracking of the concrete cover can be theoretically predicted. Guidelines for minimum concrete cover and other design criteria for concrete sea structures can be found in references 42–44. Browne and Domone[45] have summarized the results of some long-term exposure tests in the marine environment. Schaufele[46] reports on the growth or expansion of concrete sea structures from his experience along the southern California coast; he reports that concrete cylinders 13 feet in diameter have grown approximately 6 inches in diameter and 4 inches in height. Reference number 47 contains many valuable papers covering all aspects of the performance of concrete in the marine environment. Concrete should perhaps be used with caution in cold climates with large tide ranges because of the freeze–thaw cycling on the rising and falling tides and the probability of abrasion damage due to ice being moved by wind, wave, and currents. Above and below the tidal zone concrete should be quite serviceable in all climates. As a matter of material durability alone, concrete would be the material of choice for marine structures located in the tropics where steel suffers heavy corrosion and timber suffers from borer attack. Marine concrete structures properly designed and constructed should have useful design lives of from 25 to 50 years or possibly more, depending upon the location and exposure.

REFERENCES

1. Van Haagen, R. H. "Oceanographic Instrumentation," in *Handbook of Ocean and Underwater Engineering*, edited by J. J. Myers, C. H. Holm, and R. F. McAllister, McGraw-Hill, New York, 1969.
2. U. S. Navy. *Instruction Manual for Obtaining Oceanographic Data*, USNOO, H. O. #601, 1968.
3. U. S. Navy. *Marine Biology Operational Handbook*, NAVDOCKS, MO-311, 1965.
4. *United States Coast Pilot*, 10 vols. covering all U. S. coastal waters, U. S. Dept. of Comm., NOAA, issued periodically.
5. Blumberg, R., and Rigg, A. M. "Hydrodynamic Drag at Supercritical Reynolds Numbers," *Am. Soc. of Mech. Eng'rs.*, Petroleum Session, 1961.
6. Heaf, H. J. "The Effects of Marine Growth on the Performance of Fixed Offshore Platforms in the North Sea," *OTC*, paper #3386, May 1979.
7. Gosner, K. L. *A Field Guide to the Atlantic Seashore*, Peterson Field Guide Series, Houghton Mifflin Co., Boston, 1979.
8. "Marine Fouling and Its Prevention," Woods Hole Oceanographic Inst. report to U. S. Navy, 1952.
9. ZoBell, C. E. *Marine Microbiology*, Chronica Botanica Co., Waltham, Mass., 1946.
10. Muraoka, J. *The Effects of Marine Organisms on Engineering Materials for Deep Ocean Use*, U. S. Navy, Civil Eng'g. Lab., Port Hueneme, Calif., 1962.
11. Efird, K. D. "The Interrelation of Corrosion and Fouling of Metals in Seawater," Nat. Assoc. of Corrosion Eng'rs., Corrosion Conf., Toronto, 1975.

12. Benson, P. H., Brining, D. L. and Perrin, D. W. "Marine Fouling and Its Prevention," *Marine Technology*, SNAME, Jan. 1973.
13. Anonymous article, *Ocean Industry*, July 1978.
14. "Report of Marine Borer Research Committee, New York Harbor," New York, Dec. 1946.
15. U. S. Navy. *World Atlas of Coastal Biological Fouling*, USNOO, I.R. #70-51, Sept. 1970.
16. Chellis, R. D. *Pile Foundations*, McGraw-Hill, New York, 1961.
17. Turner, R. D. *A Survey and Illustrated Catalog of the Teredinidae*, Harvard University, 1966.
18. Menzies, R. J. and Turner, R. D. "The Distribution and Importance of Marine Borers in the U. S.," Symposium on Wood for Marine Use and Its Protection from Marine Organisms, Am. Soc. for Testing Materials, STP-200, 1956.
19. U. S. Navy. *Harbor Reports on Marine Borer Activity*, NAVDOCKS, P-43, June 1950.
20. *Progress Reports on Marine Borer Activity in Test Boards*, W. F. Clapp Laboratories, under B.U.Y.D. Conf. #NBy–17810, 13 issues, annually, 1946 through 1959.
21. *Tidelines – A Current Review of Pacific Coast Marine Borers*, bulletin of B. C. Research, Ltd., issued monthly.
22. Crandall, P. S. "Timber in Waterfront Construction," *Journal Boston Soc. Section, ASCE*, Vol. 54, #2, April 1967.
23. Uhlig, H. H., ed. *The Corrosion Handbook*, Wiley, New York, 1948.
24. Evans, U. R. *An Introduction to Metallic Corrosion*, Edward Arnold, London, 1963.
25. Rogers, T. H. *The Marine Corrosion Handbook*, McGraw-Hill, New York, 1960.
26. LaQue, F. L. *Marine Corrosion*, Wiley, New York, 1975.
27. Tuthill, A. H., and Schillmoller, C. M. *Guidelines for Selection of Marine Materials*, International Nickel Co., New York, 1966.
28. Ayers, J. R. and Stokes, R. C. "Corrosion of Steel Piles in Salt Water," *Proc. ASCE*, WW-3, Vol. 87, Aug. 1961.
29. U. S. Navy. *Design Manual – Waterfront Operational Facilities*, NAFVAC, DM-25, Oct. 1971.
30. Rayner, A. C. "Life of Steel Sheet Pile Structures in Atlantic Coastal States," *Proc.* 3rd Conf. on Coastal Eng'g., Council on Wave Research, Berkeley, 1952.
31. Creamer, E. V. "Splash Zone Protection of Marine Structures," *Proc. OTC*, paper No. 1274, Houston, April 1970.
32. Hedborg, C. E. "Corrosion in the Offshore Environment," *Proc. OTC*, paper No. 1958, Houston, May 1974.
33. Hanson, H. R. and Hurst, P. C. "Corrosion Control – Offshore Platforms," *Proc. OTC*, paper No. 1042, April 1969.

34. Munse, W. H. *Fatigue of Welded Steel Structures*, Welding Research Council, New York, 1964.

35. "Mariner Steel Sheet and H-Piling," brochure of United States Steel Corp., Pittsburg, Pa., 1964.

36. "Results of Eight Year Corrosion Tests Conducted by the U. S. Naval Research Laboratory on Ten Structural Steels and Wrought Iron," Engineering Service Dept. Report, A. M. Byers Co., Ambridge, Pa., 1967.

37. U. S. Navy. *Corrosion of Materials in Hydrospace*, NAVFAC, USNCEL, Port Hueneme, Calif., Dec. 1966.

38. Johnson, S. *Deterioration, Maintenance, and Repair of Structures*, McGraw-Hill, New York, 1965.

39. Kennedy, T. B., and Mather, K. "Correlation Between Laboratory Accelerated Freezing and Thawing and Weathering at Treat Island, Maine," *J. Am. Concrete Inst.* 25(2): 141, 1953.

40. Gerwick, B. C. "Marine Concrete," in *Handbook of Ocean and Underwater Eng'g.*, edited by Myers, Holm, and McAllister, McGraw-Hill, New York, 1969.

41. Bazant, Z. P., "Physical Model for Steel Corrosion in Concrete Sea Structures, Theory and Application," *Proc. ASCE*, ST-6, Vol. 105, June 1979.

42. American Concrete Institute (ACI), "Guide for the Design and Construction of Fixed Offshore Concrete Structures," ACI Comm. 357, title No. 75-72, Dec. 1978.

43. Det Norske Veritas, "Rules for the Design Construction and Inspection of Fixed Offshore Structures," Oslo, 1974.

44. Federation de la Precontrainte, "Recommendations for the Design and Construction of Concrete Sea Structures," France, 1973.

45. Browne, R. D. and Domone, P. L. J., *The Long Term Performance of Concrete in the Marine Environment, Offshore Structures*, Proc. of the Institute of Civil Engineers Specialty Conference, London, 1975.

46. Schaufele, H. J. "Erosion and Corrosion on Marine Structures, Elwood, Cal.," *Proc.* First Conf. on Coastal Eng'g., Council on Wave Research, Berkeley, 1951.

47. American Concrete Institute, "Performance of Concrete in Marine Environment," SP-65, ACI, Detroit, 1980.

Appendix 1
Conversion Factors

U. S. Customary		SI Metric
1 inch (in.)	=	2.54 centimeters (cm)
1 foot (ft)	=	.3048 meter (M)
1 yard (yd)	=	.9144 meter (M)
1 statute mile	=	1.609 kilometers (kM)
1 square inch (sq in.)	=	6.45 square centimeters (sq cm)
1 square foot (sq ft)	=	.093 square meter (sq M)
1 cubic inch (cu in.)	=	16.39 cubic centimeters (cc)
1 cubic foot (cu ft)	=	.0283 cubic meter (cm)
1 pound (lb)	=	.453 kilogram (kg)
1 ton	=	.907 metric ton (MT)
1 pound (force)	=	4.448 Newtons (N)
1 kilopound (kip) (force)	=	4.448 kilo Newtons (kN)
1 kip per foot (k/ft)	=	14.59 kilo Newtons per meter (kN/M)
1 pound per square inch (psi)	=	6.89 kilo Newtons per square meter (kN/M^2)
1 pound per square foot (psf)	=	47.88 kilo Newtons per square meter (kN/M^2)
1 pound per cubic foot (pcf)	=	16.02 kilograms per cubic meter* (kg/M^3)

* (not an S.I. unit)

Nautical Units

1 nautical mile = 1.151 statute miles = 6076.1 feet = 1.852 kilometers
1 knot = 1.151 miles per hour = 1.688 feet per second = .515 meter per second
1 fathom = 6.0 feet = 1.829 meters
1 long ton = 1.12 short tons = 2240 pounds = 1.016 metric tons

Appendix 2
Some Physical Properties of Seawater

Seawater at atmospheric pressure and salinity of 35‰, corresponding values for freshwater shown in parentheses

Temperature (°F)	Specific weight γ(pcf)	Density ρ (slugs/ft³)	Kinematic viscosity ν, $\times 10^{-5}$ (ft²/sec)	Bulk modulus of elasticity $\times 10^3$ (psi)
32	64.18	1.995	–	315
	(62.42)	(1.940)	(1.929)	(289)
40	64.15	1.994	–	323
	(62.43)	(1.940)	(1.664)	(296)
50	64.10	1.992	1.460	332
	(62.41)	(1.940)	(1.408)	(305)
60	64.03	1.990	1.264	340
	(62.37)	(1.938)	(1.211)	(312)
70	63.94	1.988	1.109	348
	(62.30)	(1.936)	(1.055)	(319)
80	63.84	1.984	0.983	354
	(62.22)	(1.934)	(0.930)	(325)

Appendix 3
Selected Periodical
Information Sources

Magazines

Ocean Industry, Gulf Publishing Co., Houston, Tex., monthly
Offshore, Petroleum Publishing Co., Tulsa, Okla., monthly
The Dock and Harbour Authority, Foxlow Publishing Co., London, monthly
World Dredging and Marine Construction, Symcon Publishing Co., San Pedro, Calif., monthly
Sea Technology, Compass Publications, Arlington, Va., monthly

Journals

Journal of Waterway Port Coastal and Ocean Engineering Division, Proc. ASCE, quarterly, New York
Journal of the Marine Technology Society, Washington, D.C., 10/year
Journal of Geophysical Research, Am. Geophysical Union, Washington, D.C., bi-monthly
Bulletin Permanent International Association of Navigation Congresses, Brussels, Belgium, quarterly
Marine Technology, Journal of SNAME, New York, quarterly
Journal of Marine Research, Bingham Oceanographic Lab., Yale University, New Haven, Conn., 3/year
Deep Sea Research, Pergamon Press, New York, 6/year
Coastal Engineering, Elsevier Scientific Publishing Co., Amsterdam, The Netherlands, quarterly

Conference Proceedings

Conference on Coastal Engineering, ASCE, 1st 1950, 15th 1976
Civil Engineering in the Oceans, ASCE, intermittently
Offshore Technology Conference, Houston, Texas, annually since 1969
World Dredging Conference, World Dredging Assoc., every 18 months since 1967
International Conference on Port and Ocean Engineering under Arctic Conditions, biannually since 1971

Universities and Government Organizations

Many universities, especially those participating in NOAA, sea grant programs, publish newsletters and reports on marine research and engineering. U.S. government agencies, such as NOAA, the U.S. Army Corp. of Engineers, and the U.S. Naval Oceanographic Office, and Facilities Engineering Command, produce a tremendous volume of technical reports and bulletins through their many internal branches.

Index

Index